TERRELL'S TEXAS CAVALRY

By John W. Spencer

EAKIN PRESS ★ BURNET, TEXAS

FIRST EDITION

Published in the United States of America
By Eakin Press, P. O. Drawer AG, Burnet, Texas 78611

ISBN 0-89015-300-0

ACKNOWLEDGMENTS

It would be impossible to acknowledge all the courtesies and assistance received in the course of writing this history. But imperfect as it is, some attempt must be made to thank as many of the persons and institutions as possible.

My first thanks go to my wife, Betty, for laboring over my handwriting, and the many hours of typing the manuscript. I would especially like to thank the people in the Texas and Genealogy section of the Dallas Public Library from where much of the material in this book was researched. They were very helpful in locating references for me, in addition to providing the facilities for microfilm reading. The National Archives in Washington, D.C., was also very helpful in providing me with a microfilm of letters and documents from the files of Terrell's regiment stored there. The Eugene C. Barker Texas History Center, University of Texas at Austin, was also very helpful in providing access to information in the files of the history center.

Other research was done at the Archives Division of the Texas State Library, the Corsicana Public Library, Corsicana, Texas, the W. Walworth Harrison Public Library, Greenville, Texas, the New Orleans Public Library, New Orleans, Louisiana, and the Confederate Research Center, Hill County Junior College, Hillsboro, Texas.

The people of the Mansfield Battlefield Park and Museum were very courteous and informative in the research on the Battle of Mansfield, as were the local citizens of Pleasant Hill, Louisiana with regard to the battle fought near their city. The people around Cayuga, Texas, especially the Vaughn family, were very helpful in locating the old Wildcat ferry crossing where three of the regiments would be disbanded at the end of the war. The Vaughn family was very generous in allowing me access to the Trinity River which adjoins their property and assisting me in location the actual ferry crossing.

Two young ladies were very helpful in copying names from the microfilm for the roster of Terrell's regiment. They were my good friend, Miss Risa Rasmussen, and Miss Beverly Spencer, my daughter.

TABLE OF CONTENTS

Acknowledgments iii
Preface v
CHAPTER I 1
First Regimental Action
Organization of the Regiment
Mutiny
CHAPTER II 14
The Delaying Action and the Battle of Mansfield
CHAPTER III 31
The Battle of Pleasant Hill
The Battle of Blair's Landing
CHAPTER IV 45
Skirmishes After Blair's Landing
Monett's Ferry
Battle of Lecompte
Regimental Move to Marksville
Battle of Mansura
Battle of Yellow Bayou
Duty After Yellow Bayou
Skirmishes At Morgan's Ferry
Terrell Given Brigade
Fight at Morgan's Ferry
Fight on the Atchafalaya
Duty on the Bayou Beouf
Winter at Alexandria
Likens' Regiment Transferred to Texas
Furloughs
CHAPTER V 58
Return to Texas
Disbandment at Wild Cat Bluff
Back to Shreveport
Terrell Promoted to Brigadier General
Terrell Goes on to Austin
Terrell and other Officers Go To Mexico
Terrell in Monterrey
To Mexico City
Maximilian
Decision to Return to Texas

APPENDIX I 92
Alexander Watkins Terrell
APPENDIX II 101
Sketches of the Companies
Regimental Strength
Company Commanders
Field and Staff
APPENDIX III 110
Roster of Terrell's Regiment
Gray's Company
Mullins' Company
APPENDIX IV 158
Likens' Regiment
APPENDIX V 166
Yager's Regiment
APPENDIX VI 171
Herbert's Regiment
APPENDIX VII 178
Roster of Herbert's Regiment
APPENDIX VIII 183
Surrender Terms of the Trans-Mississippi Department
APPENDIX IX 186
Letters by Generals E. Kirby Smith and J.B. Magruder
BIBLIOGRAPHY 188
INDEX 195

PREFACE

Unfortunately for history the men of Colonel Alexander W. Terrell's Texas Cavalry Regiment were not disposed to keep personal diaries or detailed accounts of their service during the two years of the regiment's existence. There were many educated men in the regiment so it was not for lack of education or ability that they did not write. No doubt there are yet to be found letters from members of the regiment to relatives which would yield valuable information about the regiment and Texas during the period. The account of the regiment in this book was woven together from the many sources listed in the bibliography. Even Terrell, himself, never saw fit to write of his experiences with the regiment during the War, although he did write a book in his later years about his stay in Mexico immediately after the Civil War. Not much can be found in it about the regiment, however, and the book was written some thirty-five to forty-five years after the war ended.

Terrell's regiment was organized from several unattached companies of men, mainly from the eastern part of the state. Organization was first into a battalion, later into the regiment. Many of the men, especially the officers, had seen action in the Civil War in other areas and for various reasons found themselves in the unattached companies which formed the nucleus of the regiment. The regiment was organized as the Thirty-fourth Texas Cavalry Regiment and this is the only numerical designation ever used by Terrell's men. Texas regiments were designated with the next available numerical name according to the sequence of organization.

Terrell's regiment was organized in mid-1863 and, therefore, rightfully should have had a numerical designation lower then either Likens' regiment or Brown's regiment, both of which were organized in late 1863 and both of which claimed the "Thirty-fifth" as their regimental designation. Terrell's regiment claimed the "Thirty-fourth" as their number, although another regiment had been so designated in early 1862. This regiment, however, was "dismounted" permanently in late 1862 and fought as infantry for the remainder of the war, retaining the "Thirty-fourth," but adding (Dismounted) as a part of their name. The National Archives in Washington, D.C., shows Terrell's regiment as the "Thirty-seventh" Texas Cavalry Regiment, although it was organized before both the Thirty-fifth and the Thirty-sixth. The men of Terrell's regiment always referred to their organization as the "Thirty-fourth," or sometimes simply "Terrell's" Texas Cavalry Regiment. The regiment was one of the few remaining mounted regiments at the end of the war and the men all signed their paroles as the "Thirty-fourth" Texas Cavalry Regiment.

February 29, 1864, was the date of the last and only known muster roll of the ten companies in the regiment. These are in the National Archives in Washington, D.C., probably having been surrendered to the Union Army in Shreveport, Louisiana, or possibly Houston, Texas. If any muster rolls were submitted prior to that date, they were either lost or destroyed. Probably none were made after February 29, 1864, because of the wild excitement and grinding activity connected with the Red River Campaign in Louisiana.

Terrell's Texas Cavalry Regiment is only one of a number of military organizations which served Texas and the Confederacy faithfully and courageously for a short period of time, and which have received very little attention from the writers of history. That the men of the regiment did not write of their experiences is unfortunate because their suffering and hardships were just as great as those who fought east of the Mississippi River. Although the battles and skirmishes may not have been on the same scale from the standpoint of the number of men involved in them, the gallant charges they made were just as heroic, the wounds they received were just as painful and disabling, and death was just as final. As C.H. Jackson stated in *Texans Who Wore the Gray*, "Many of our Comrades east of the river, claim that we Texans on this side were 'not in it,' which fact must be inferred from the very meager accounts published in the news of the day as to what took place on this side of the Mississippi; and as there seems to be but little known or said about the chivalrous manner in which Gen. Taylor, with his brave following, turned the tide of battle at Mansfield and

Pleasant Hill; which saved the Western Department, and especially Texas from the most sad fate, that was the misfortune of any part of our beloved country to have suffered.''

The Thirty-fourth Texas Cavalry Regiment was not the only regiment commanded by Terrell. In Louisiana Terrell was frequently made commander of a brigade of Texas cavalry consisting of the Thirty-fourth, Thirty-fifth, and 1st regiments of cavalry. The Thirty-fifth was commanded by Colonel James B. Likens, who had been with his battalion as early as November, 1861, at Sabine Pass, Texas; the 1st cavalry was commanded by Colonel William O. Yager, who assumed command and was promoted to colonel after the death of Colonel Augustus Buchel at the Battle of Pleasant Hill, Louisiana, on April 9, 1864. In February, 1865, Likens' regiment was dismounted and returned to Texas from Louisiana, where Terrell's brigade was a part of a division of cavalry composed of his own brigade and one or two brigades of Louisiana cavalry. Replacing Likens' regiment in Terrell's brigade in Louisiana was Bagby's old 7th Texas Cavalry, which he had commanded and had been with during the Sibley campaign to New Mexico in late 1861 and 1862, arriving back in San Antonio in May, 1862. Command of the 7th Cavalry was now in the hands of Colonel P.T. Herbert or possibly Colonel Gustave Hoffman.

Since those three additional regiments were also a part of Terrell's cavalry, brief histories of each are included.

Three of the regiments were disbanded at Wild Cat Bluff on the Trinity River, near the present community of Cayuga, Texas, these being the Thirty-fourth, 1st, and 7th Texas Cavalry regiments. The other regiment, the Thirty-fifth, was not disbanded until several days later near Houston, Texas, having been transferred to Galveston.

Many books and manuscripts have been written about other more publicized Texas military units of the Civil War period, but very little had been written about the units which served in Texas and in the Red River Campaign in Louisiana. It is hoped that this book will help to fill that void.

John W. Spencer

TERRELL in Mexico, 1865.

1

First Regimental Action

Cannons could be heard booming through the tall virgin pines south of Mansfield, Louisiana, as Terrell's Texas Cavalry Regiment trotted in column of twos toward Ten Mile Bayou. The day was cloudy and an occasional shower settled the swirling, gritty cloud of dust which surrounded the gray lines of horsemen. Fighting had been going on since early morning and now ambulances carrying the wounded bounced along with other wagons and grim-faced soliders marching northward toward Mansfield.[1]

General Richard Taylor[2] was waiting for the cavalrymen at Six Mile Bayou where the regiment stopped to await his further orders. Debray's 26th Texas Cavalry and Buchel's First Texas Cavalry were with Terrell's regiment, all a part of a division under the immediate command of General H.P. Bee.[3] Both Debray's[4] and Buchel's regiments were well-disciplined cavalry units, armed with sabers;[5] and Colonel Buchel,[6] a former Prussian soldier, had drilled his men from the New Braunfels area of Texas in the European fashion.

An hour and a half passed before General Taylor rode off for Mansfield and three cavalry regiments remounted and galloped on to report to General Tom Green[7] at Ten Mile Bayou where heavy skirmishing was taking place.[8] In less than an hour the sharp odor of cannon powder and musketry smoke lay heavy on sweaty horses, and the sound of squeaking saddles, bumping canteens, cartridge boxes, and rifles was drowned out by the booming of cannons and rattle of muskets firing in rapid succession. Buchel's regiment wheeled about into battle position on the left of the battle line, along with Lane's brigade, which was already there. Debray's regiment was formed into position for a charge, and Terrell's regiment moved back on the road and waited in reserve, ready to charge to any position they might be needed. Brisk skirmishing continued on the right of the line and artillery continued with grape and canister until dark;[9] and then Terrell's cavalrymen pulled their oil blankets over them and lay with muskets in their arms, not knowing what the night would bring. Although the wild horsemen of Texas still had not been tested in a fight, the time was near as Yankees were just on the other side of the creek. Rain poured down on them now and the dirty, tired men talked about the possibilities of a battle and tried to sleep.

First Battle

Before dawn the next morning on April 8, 1864, Terrell's, Debray's, and Buchel's regiments were left to face the enemy alone as General Green and other Confederate troops rushed back toward Mansfield to prepare for battle.[10] The months of drilling and training would soon be put to the test of Yankee muskets and canister and whether Terrell's cavalrymen would be a disciplined fighting regiment or the same raw farmers and merchants of a few months before would soon be known.

Organization of the Regiment

Since June, 1863, when the regiment was organized,[11] these men had drilled daily under Colonel Terrell, Lieutenant Colonel Robertson,[12] and Major Morgan.[13] Most of the company commanders and their men came from East Texas areas, where they had been organized in late 1862 or early 1863 as independent companies. Colonel Terrell had combined some of them, first into a battalion and later into a regiment, initially to fight the Indian raiders out of Oklahoma Territory and Kansas, who were ravaging the people of North Texas.[14] Inevitably, as it is in all armies, nothing ever goes exactly as it is planned and the regiment never made it to their first duty station at Bonham where they were ordered shortly after organization on May 30, 1863.[15] Apparently it was felt that they should be sent to Camp Groce near Hempstead, Texas, for more organization and training before going on patrol or on any campaign against the Indians.[16] On July 8, 1863, the regiment was ordered again to Bonham, where Terrell was to take command of the Northern Sub-District in the absence of Brigadier General Bankhead.[17] Arms would be supplied to the regiment, but at this point there were only the personal arms brought by the men when they enlisted.

Company B was ordered to Waco and vicinity to break up nests of deserters and the remainder of the regiment was ordered to the mouth of the Brazos River near Velasco to defend against an expected invasion by the Federals.[18] Part of the regiment was dismounted and ordered to Galveston Island on about August 22, 1863, to help quell a mutiny there by some of the infantry troops[19] and Gould's regiment was remounted and replaced Terrell's in going to Bonham.[20] On August 27, 1863, the regiment was ordered to take post at Columbus or near West Bernard, as felt best by Colonel Terrell.[21]

General E. Kirby Smith[22] wanted to place Colonel Terrell in charge of the Cotton Bureau in September, 1863, but Terrell would not accept such position, although he was a member of the investigating committee and reported to Houston for that purpose in the same month.[23] While in Houston command of the regiment was left with Lieutenant Colonel Robertson, who was a very capable officer, but was to have the unfortunate experience of a small mutiny by several members of the regiment.

While at Camp Kelsoe Springs in Colorado County on September 11 Colonel Robertson received orders late in the evening to leave the horses and equipment with a detail of men and proceed with the regiment to Galveston. At this time there were about 340 men present in the regiment as some of them were already on duty in Galveston and Company B was still on detached service. The regiment was paraded into formation and the orders from General Magruder were read to them along with detailed instructions for the march to Galveston.

Captain Murray, commanding Company I, addressed Colonel Robertson when the Colonel's instructions had been completed, saying that this looked like a permanent dismounting to him and that he had promised his men when they enlisted that they would not be dismounted, and that he was not willing to allow this to happen. The troops frequently cheered Captain Murray during this time and the excitement became so intense that Colonel Robertson was unable to suppress it. Still unarmed, he would have been powerless to arrest the men, but he addressed them again and tried to convince them that Captain Murray was misleading them. He was frequently interrupted by cheering for Captain Murray, and with the excitement increasing every moment and it becoming obvious that he could no longer control the formation, Colonel Robertson dismissed the regiment, hoping to have better control of the situation the following morning.

On the morning of September 12, Colonel Robertson rode up the camp lines and found a large number of the men saddled up and preparing to leave. He addressed them again and offered to call for volunteers to go to Galveston instead of dismounting the entire regiment for this purpose. This generally seemed to satisfy them and he retired to his quarters only to be informed shortly thereafter that some of the men had ridden out of camp, taking with them many of the horses, baggage, and clothing of the men who had previously been sent to Galveston. The troops left in a body with Captain Murray leading them, and consisting of twenty-five men from his own company, four men from Captain Gray's company, thirty men from Captain Spike's company, and thirty men from Captain Hurley's company. Lieutenant J. G. Chancellor, who had also been active in the events preceding the mutiny, went with them.[24]

On September 30 nine of Terrell's company commanders addressed a letter to the commanding general concerning the mutiny. In it these commanders reviewed the original organization of the regiment as part of the Arizona Brigade, contemplating an active cavalry service, and how the men of the regiment were those who had seen cavalry service during the war with the North and had worked hard to become efficient cavalry units, approaching the highest state of efficiency when dismounted. Their horses had been purchased with great expense and their equipment was the most approved style for cavalry service.

These things were presented as extenuating circumstances for the mutiny with a promise that if the regiment were remounted at once, the regiment would become unsurpassed by any other in the service. The letter assured the general that the men who deserted the ranks were loyal and had only deserted to join some other cavalry regiment, and who would return to the regiment if it was remounted and put into active field service.[25]

There is no record of any reply or response by the commanding general to the letter by the company commanders. However, Colonel Terrell lost no time in sending a detachment under Lieutenant Colonel Robertson to pursue and bring the men back to the regiment. Colonel Robertson and a detachment of twenty-five men left the camp near Columbus on October 17 in pursuit. Lieutenant Starr had apparently preceded him and was returning with a few of the men when met by Colonel Robertson near Millican. From him they learned that Captain Murray was probably near his home in the eastern part of Wood County, and later learned that he was camped with about sixty armed men near Springville in Wood County, ready for resistance. Lieutenant Colonel Robertson continues in a report he made to the commanding general on December 2. . . .

> . . . When the detachment reached Henderson County, Monday, the 26th October, the county in which Lieutenant Chancellor resides, I took 4 men, left my line of travel 10 miles, captured Lieutenant Chancellor near his home, carried him to the conscript camp near Tyler, and delivered him to Major Tucker, commanding camp, for safe-keeping. I reached Tyler the evening of the 27th, and remained there the next night, awaiting the return of the men sent in advance. That night at 10 o'clock they reported to me. They had not seen Captain Murray, but had been to his camp, found a few men there, and that Captain Murray and his men were to rendezvous there on Monday, the 26th, one day before I reached Tyler. They

stated that Captain Murray was then out hunting up his men, to report with them to Brigadier General McCulloch at Bonham. I started with the detachment the next morning at daylight. That day I met and arrested 3 of his men, who were sent out by him to bring in others. From them I learned that Captain Murray, with 8 or 10 men, had gone to the northern part of Wood County, to Mr. Gilbreth's. I sent the detachment on to his camp, and took 7 men and went to Mr. Gilbreth's that night. A little after daylight I arrested Captain Murray. He had no men with him. The party who went to his camp and the party under me who captured Captain Murray marched 75 miles during the day and night. With Captain Murray I proceeded to Quitman, where, by previous order, I met the detachment with prisoners, amounting in all to 15. With only 25 men, I deemed it prudent to place my prisoners in the custody of Major Tucker, commanding camp near Tyler. I turned them over to him; remained one day to rest horses and men, for both had been on duty night and day most of the time we were in Wood County. I started on a scout into Van Zandt and Kaufman Counties. In Van Zandt I arrested some deserters from other regiments, who had furloughs for fifteen days from Brig. Gen. H.E. McCullouch, commanding Northern Sub-District of Texas. One of these men brought me a paper showing the names of about 100 deserters who had been organized into a company. Among the names were 23 of Colonel Terrell's regiment, for whom I was then hunting. I herewith respectfully forward a list of the names belonging to Colonel Terrell's regiment, and a copy of the official certificate on the back of the paper containing them, marked A.

I found that most of these men of Colonel Terrell's regiment had gone to Bonham; others had taken to the woods. My trip to Van Zandt and Kaufman proving almost fruitless as to arresting men of this regiment, I arrested all I found of others, and carried them to Tyler. Satisfied that some arrangement had been made with deserters, I dispatched a courier to Brigadier General McCulloch for information, furnishing him with a copy of my orders from Major General Magruder. I returned to Tyler with a number of prisoners from this and other regiments, and

> there received the answer of Brigadier General McCulloch, which I respectfully forward herewith, marked B.
>
> On my arrival at Tyler, I found an order from Major General Magruder directing me to proceed at once with my detachment to my regiment. Not knowing where the regiment was, I came with the detachment to Houston, and this day have turned over Captain Murray, Lieutenant Chancellor, and 23 privates to Major Hyllested, provost-marshal general.[26]

Captain Murray and Lieutenant Chancellor remained in prison, but the men arrested with them were given a military trial, cleared, and returned to the regiment. This trouble caused Terrell to request to be relieved of responsibility for the Cotton Bureau, but General Smith wanted another investigation and Terrell reluctantly agreed and made his report on March 3, 1864. This was his last duty with the Cotton Bureau and after this, he gave his entire attention to the regiment.[27]

A list of troops in the District of Texas, New Mexico, and Arizona under Major General J.B. Magruder, C.S. Army, on September 30, 1863, shows nine companies of Terrell's regiment to be in the Eastern Sub-District under Col. X.B. Debray on duty in the defense of Galveston.[28] Company B was still on duty in the Waco area. Most of the month of October was spent by Lieutenant Colonel Robertson and twenty-five men in pursuit of the deserters of September.[29] The remainder of the regiment moved to camp near Columbus, Texas, where they had been ordered. In November, 1863, Terrell's regiment was ordered toward Matagorda and to go on to Lavaca via Texana.[30] Terrell's regiment was part of a Confederate force of about 6,000 men which were being sent out to oppose a Federal force that had landed at the mouth of the Rio Grande and which retreated before reaching the Confederate force.[31]

In December, 1863, the regiment was at Camp Wharton, Texas, and on December 23, 1863, was camped at Jones Creek near the Gulf.[32] On December 31, 1863, the regiment is shown as at "Gulf Prairie" along with Buchel's cavalry regiment, a cavalry battalion under Lieutenant Colonel R.R. Brown, and two batteries of artillery.[33] An abstract from the morning report of the First Brigade, Second Division, on January 6, 1864, at Camp Wharton shows Terrell's regiment with twenty-five officers and 402 men present for duty. The aggregate present and absent was 701, which included Company B still on detached service chasing deserters.[34]

On January 10, 1864, Company B stationed near Waco received orders to report to the regiment. The company marched 200 miles and joined the regiment at Camp Dixie, Texas, on January 21, 1864.[35]

From Camp Dixie the regiment marched about twenty-five miles to Camp Sidney Johnston near the mouth of the Caney River on January 22, making the trip in thirty-three hours. On January 26 they marched back to Camp Dixie and then back to Sidney Johnston on February 2. Remaining there until February 19, they then marched to Columbia, took the railway to Houston on February 21, arriving the same day, and on the following day took another railway for Virginia Point, arriving the same evening. The regiment then passed over the railway to Galveston Island on February 24, 1864. The horses in the meantime were sent to Fayetteville, Texas, along with the necessary detachment to care for them.

During this time the court-martial trials of the deserters were held at Camp Sidney Johnston, probably taking only a few days. On February 24, 1864, S.O. No. 55 ordered Terrell's regiment to take post at Galveston.[36] On February 28 General Magruder wrote to General Bee asking how long it would take to remount the troops of Terrell's regiment.[37] A return of the 3rd Brigade, 2nd Division, for February, 1864, shows Terrell's regiment, with all ten companies present for the first time since August of the preceding year. Present for duty were thirty-two officers and 403 enlisted men. Total present and absent was 734.[38] The horses of Terrell's regiment had been sent to Fayetteville in Fayette County with a detachment of the regiment and were there on March 1, 1864.[39] S.O. No. 72 on March 12, 1864, ordered Brigadier General H.P. Bee to proceed without delay with his whole available force of cavalry and artillery to Alexandria, Louisiana, reporting to General Taylor.[40] The same day the Trans-Mississippi Department in Louisiana was writing Major

General Magruder to send Terrell's and Buchel's regiments to General Taylor immediately, reporting to General Green.[41]

On March 17, 1864, S.O. No. 76 organized the cavalry forces which had been ordered to Louisiana. Terrell's regiment was brigaded with Likens'[42] and Woods'[43] regiments under General Bee. Buchel's,[44] and Debray's[45] and Gould's[46] regiments were formed into a brigade under Brigadier General Howes.[47] While Terrell was remounting his regiment at Rusk, Texas, he received orders from the Trans-Mississippi Department to move on to Natchitoches, Louisiana, sending couriers to General Taylor at that point, notifying him of the regiment's movements.[48] This order was changed and the regiment was ordered to report to General Taylor at Pleasant Hill, Louisiana, by the shortest and most practicable route.[49] General E. Kirby Smith sent a letter to Texas Governor P. Murrah on March 31, 1864, warning that the Federals were moving in two columns up the Red River and also across Arkansas, that he would try to stop them, but that Texas should get out the State Troops to replace those which had been sent to Louisiana because he was sure a landing would be made on the Texas coast as soon as the Federals learned of the weakness.[50]

Terrell's regiment arrived at Sabinetown on the Sabine River on April 1 and headed toward Pleasant Hill. However, further word ordered them to Mansfield, Louisiana, by way of Logansport,[51] where they reported to General Taylor on April 5, with about 360 men along with General Bee and the regiments of Debray and Buchel, each consisting of about 500 men.[52]

As the Confederate forces began concentrating in and around Mansfield, the Federal forces continued their march northward toward Shreveport. They were commanded by General Nathaniel P. Banks,[53] who had received orders to invade the interior of Louisiana, with the objective of capturing Shreveport, three hundred and fifty miles above New Orleans, the seat of the Confederate government of Louisiana, and where an immense amount of cotton was stored.

Three columns would advance toward Shreveport, General A.J. Smith was to advance from Vicksburg up the Red River with 10,000 men on loan from General Sherman and with orders that these men were to go no further than Shreveport, and that they would be returned to General Sherman by April 15.[54] General Banks would advance from New Orleans with a force of about 20,000, and General Steele would drive southward from Arkansas with his army of about 10,000. In addition a fleet of about sixty Union gunboats, ironclads, monitors, transports, and quartermaster boats would force its way up the Red River, giving support to the ground troops.[55]

Cotton was the real objective of the entire Red River expedition. Textile mills in General Banks' home state of Massachusetts, and the other states in the Northeast were almost out of business for the want of it. As early as October, 1862, a delegation of Bostonians, representing all New England manufacturers, with the support of the governor of Massachusetts, made a special trip to Washington to press the government for a quick occupation of Texas in order to obtain a supply of cotton. The capture of Shreveport was to be a prelude to the invasion of East Texas, where the rich fields would supply the raw material for the starved industry of the Northeast.[56]

Orders had been issued for the invasion of Texas, and General Banks later recalled that General-in-Chief Henry W. Halleck had sort of a "fetish" about such invasion, and had written to Banks on August 10, 1863, saying, "If it be necessary, as urged by Mr. Seward, that the flag be restored to some one point in Texas, that can be best and most safely effected by a combined military and naval movement up Red River to Alexandria, Natchitoches, or Shreveport, and military occupation of Northern Texas."[57]

Fort deRussy was manned by about 300 Confederate troops, and was the first to fall to A.J. Smith's men on March 14. On the afternoon of March 15 the leading boats of the Federal fleet reached Alexandria and found that the Confederates had evacuated that same morning. On March 31 Union cavalry occupied Natchitoches, and on April 1 the head of the Federal force reached that city. General Banks arrived at Grand Ecore about four miles from Natchitoches on April 3 aboard his headquarters boat, the *Black Hawk*.[58]

Up to Grand Ecore General Banks had been able to travel

on a road within easy communication of his boats on the Red River. At Grand Ecore, however, the Shreveport road veered off to the west away from the river and the protection of the boats. Banks had to decide whether to follow this road, send out a reconnaissance to find out if there was another road close to the river, or cross the river in hopes of finding another road on that side which would keep them in close communication with the boats on the river. In the interest of time, he decided to follow the Shreveport road up through Pleasant Hill and Mansfield, although to do so would mean that he would not be in close communication, or have the protection and supply of the fleet on the river again until he reached Shreveport. In fairness to Banks it should be noted that he did not know of a road on the west bank of the river, and most surely would have followed it if he had. Had he done so the entire campaign might very well have been successful. The Federals would have had the river with a mighty fleet of some 210 guns on their right flank and a force of some 30,000 men and 90 cannons, which should have been able to roll over any resistance which could be put in front of it by the Confederates.[59]

In the meantime, on April 2 Debray's Texas cavalry regiment was enroute as ordered to report to General Taylor at Pleasant Hill. At a short distance from Many, Louisiana, the regiment received an order to take the road to old Fort Jessup and join Colonel Bagby's brigade of Texas cavalry, on outpost duty, leaving the wagons to follow the Pleasant Hill road. Ammunition was issued to the troops and the regiment started out on the road through a dense rolling pine forest. They heard artillery fire ahead, and increased the gait of their horses, soon hearing the crackling of musketry. It was not long till they discovered that Debray's regiment had come up on the rear of a Federal cavalry force which was engaged with a force of Confederates. Debray's regiment was deployed in battle line and skirmishers were sent forward to make contact with the Federals. The Yankees were at first confused by the presence of Debray's regiment, but soon turned against them, and at this time three firings were distinctly heard — Bagby's, the Federals', and Debray's. After a short time Bagby's firing was no longer heard and the Federals were able to concentrate more of their fire on Debray's regiment. At about this time Colonel Hoffman of Bagby's brigade and Captain Corwin of the staff of Green's brigade came by a circuitous route to inform Debray that Bagby had to fall back because he had exhausted his ammunition. They indicated that the Federal force was a division of cavalry and mounted infantry, and that Debray, too must fall back to avoid being cut to pieces or captured. Debray's regiment then fell back coolly and in perfect order, keeping the Federals in check.[60]

Banks' army set out from Grand Ecore on April 6,[61] and by April 7 the head of the infantry reached Pleasant Hill, with the cavalry continuing to advance, probing for the Confederates.[62] Few precautions were taken by the Federal army as they were sure the Rebels would not make a stand before Shreveport, if then. This is indicated in a letter from Banks to General Halleck on April 2, when he stated, "General A.J. Smith, with a column of 10,000 men is with us. Our troops occupy Natchitoches, and we hope to be in Shreveport by the 10th of April. I do not fear concentration of the enemy at that point. My fear is that they may not be willing to meet us there. I shall pursue the enemy into the interior of Texas for the sole purpose of destroying or dispersing his forces if it be in my power . . . Taylor's forces are said to be on that line (Sabine town). This will not divert us from our movement."[63]

Thus the Federal army moved almost leisurely up the narrow road, flanked by towering pine trees, with occasional rain the only thing making progress difficult. The cavalry was in the front of the column, followed by their own wagon train of about 300 wagons. Next came the infantry, and their train of about 700 wagons, with General A.J. Smith's 10,000 men bringing up the rear. The entire column was about twenty miles long, a fact which would be a factor in their inability to bring their entire force into action against the Confederates.[64]

General Taylor had stopped his retreat on April 3 at Mansfield, and on April 5, General E. Kirby Smith had joined him there where they discussed their strategy. General Smith had already sent Generals Churchill's and Parsons' divisions

NATHANIEL P. BANKS, Union Commander

from Shreveport to Keatchie, which would be within supporting distance of both Shreveport and Mansfield.[65]

During the afternoon of April 7 at Wilson's farm, three miles north of Pleasant Hill, the Confederates made a stand against the huge Federal force. Brisk skirmishing occurred for about two hours, with heavy losses on both sides. Finally the Rebels fell back to Carroll's mill, eight miles or so north of Pleasant Hill and set up an even stiffer resistance than at Wilson's farm.[66] The fight at Carroll's mill put a stop to the Union advance for the day, and the next morning the regiments of Terrell, Buchel, and Debray would be facing them, contesting them all the way to Mansfield, where the biggest surprise yet was in store for the unsuspecting Federals.

2

The Delaying Action and the Battle of Mansfield

General Bee's job was to delay the Federal advance, giving General Taylor time to make preparation for a stand at Mansfield. To accomplish this he formed the regiments of Terrell, Buchel, and Debray in lines of battle, 500 yards apart and dismounted. The first line fired on the Federals, forcing them to deploy. This line held its position as long as it could and then mounted and retreated behind the other two regiments, while the next line opened fire and held its position as long as possible in the same manner as the first regiment had done.

At one point the regiments of Terrell and Buchel were placed in a strong position and ambushed the Federals, killing and wounding many of them.[1] Brigadier General Albert Lee of the Federal cavalry attested to the effectiveness of the delaying action by reporting at 11:45 a.m. on April 8 five miles from Mansfield that the Confederates had disputed their progress at every favorable position, that the Federals suffered in killed and wounded, and that Lieutenant Colonel Webb of the 77th Illinois had been killed along with two or three other officers, with several wounded.[2]

Just before noon[3] and about seven hours after the delaying action had started, Terrell, Buchel, and Debray arrived back at the battlefield site selected by General Taylor near Mansfield.[4] The site was an open field 800 yards from north to south and 1200 yards in width from east to west, through the center of which passed the road to Pleasant Hill from Mansfield, about three miles to the northwest. On the north side of the field was a

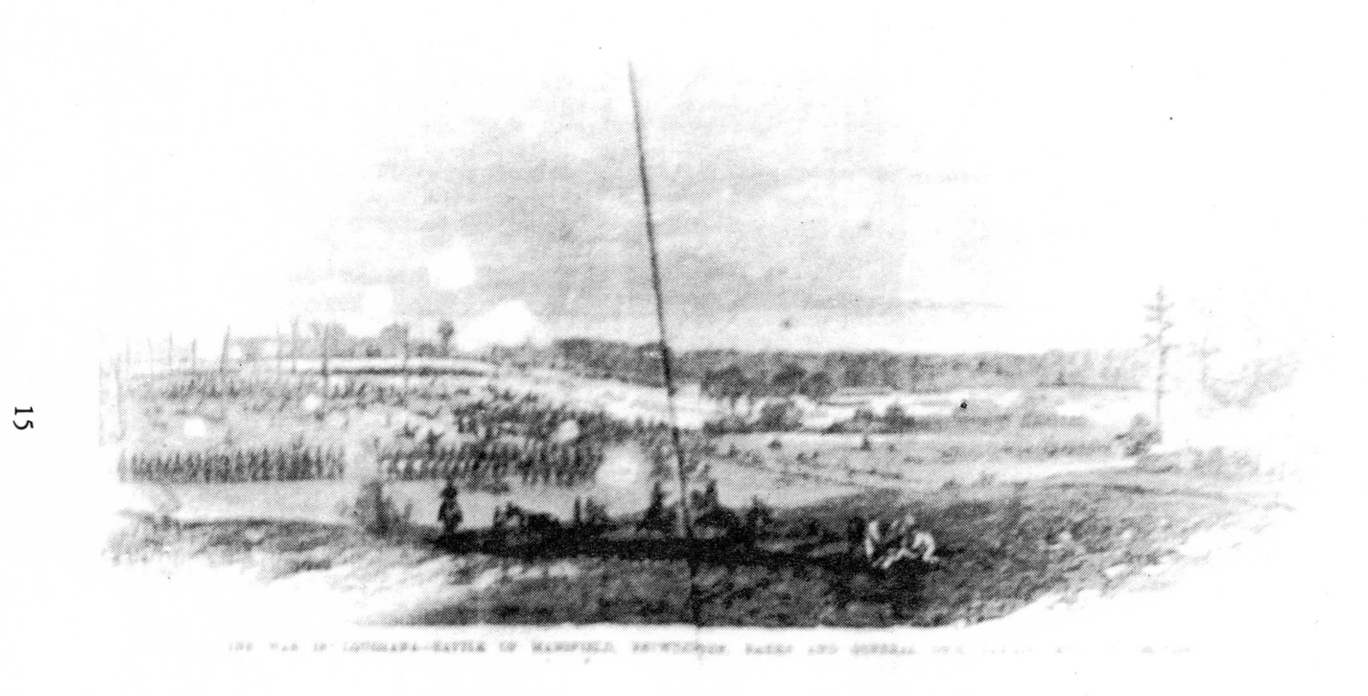

THE ONLY KNOWN sketch of the battle of Mansfield, Louisiana.

(From Leslie's Magazine, 1864.)

fence separating it from the pine forest, behind which some of the Federals would later take position. Another fence extended in a southerly direction on the west side of the battlefield, and some of the Confederates would later take their initial positions behind it. The regiments of Terrell and Buchel were placed on the extreme right of the battle line[5] and Debray's regiment was held in reserve on the main road.[6] Walker's Texas Infantry[7] was on the left of Terrell's regiment, all located on the right of the Pleasant Hill Road, which ran from the northwest to the southeast across the soon to be battlefield. Randal's Texas Infantry Brigade[8] was just on the right of the Pleasant Hill Road, facing almost in the direction the road ran. On the left of Randal's brigade on the other side of the road was Mouton's[9] Division of infantry, which included Polignac's Texas Brigade[10] and Gray's Louisiana Infantry.[11] The Valverde Battery[12] was next on the left of Gray and next to it Vincent's Louisiana Cavalry,[13] Bagby's[14] and then Lane's[15] Texas Cavalry Brigades, all part of Major's Cavalry division.[16]

At about the time Terrell's regiment came rushing in, a shower of bullets hit Mouton's line from advancing Federals, but return fire from the 18th Louisiana sent them back with heavy loss. General Taylor's horse was shot from under him during this volley from the Federals.[17]

In the early afternoon the Federals made a shift in some of their right wing troops and the Confederates, thinking that they were going to attempt to turn the Confederate left, moved Terrell's regiment from the extreme right of the line across the road to Mouton's left between Gray's Louisiana Infantry Brigade and Vincent's Louisiana Cavalry Brigade,[18] where they would fight dismounted as would the other cavalry on the left. At this point, Terrell's regiment was directly across the open field from the cannon of the First Indiana Light Artillery Battery on the Federal side, which would be firing directly into them as they charged across the field.[19] Debray's Texas Cavalry regiment was deployed along the road as a cover for the movement of Terrell's regiment, while at the same time Randal's Texas Infantry Brigade was moved from the right of the Confederate line across the road to Mouton's right, next to the road.[20] By four o'clock the battle lines had been drawn. The Confederates, numbering about 8,800 of all arms and the Federals numbering about 4,800 of all arms[21] faced each other across the open field. The Federals had other troops not far behind and would eventually have about 12,000 engaged in all three phases of the Battle of Mansfield. General Taylor sensed that the Federals were waiting for additional troops to arrive so he ordered Mouton to attack.[22]

The Louisiana Crescent Regiment led the charge across the open field, with General Mouton riding his horse in front of two gray lines of Confederates. Federal muskets from five regiments[23] and cannons opened a murderous fire and the Louisianans and Texans on their right stormed on at the quickstep. Striking first was the Louisiana Crescent Regiment, hitting the 130th Illinois while the 18th Louisiana struck the 77th Illinois, and the 28th Louisiana hit the 19th Kentucky. Polignac's Texas attacked the 77th Illinois. Terrell's regiment charged on foot alongside the 28th Louisiana, keeping pace with them[24] and Colonel Terrell leading on horseback. Randal's brigade advanced his regiments in echelon on the right of Polignac's men.

Hissing Minié balls filled the air and the smoke from the muskets and cannons darkened the sky. Bullets and grapeshot ripped through the Confederate lines, thudding into chests, tearing at arms and legs, and popping Rebel heads back in instant death. Down went the 18th Louisiana's Colonel Armant, dropped from his saddle while waving his sword. Colonel Beard of the Crescent Regiment fell dead as did Colonel Walker of the 28th Louisiana. Men from Terrell's regiment fell with the Louisianans. Down a hill and into a ravine they plunged, up the hill toward the Federals the gray lines charged on. Grape and canister and Minié balls continued to pour into them, now within 200 yards.[25] The thunder of cannons and the continual roar of the muskets drowned out the screams of the charging Confederates, many now lying bloody on the field. Three Minié balls knocked General Mouton off his horse.[26] And the Confederate charge was checked. Mouton's men laid down and returned the Federal fire.[27] But General Green mounted some of the cavalry and rode over to help.[28] And in the meantime, the 34th Texas Cavalry (Dismounted), who had been left to support the artillery, rushed over the open field. Led by Lieutenant Colonel Caudle they thrust forward with the bayonet on the Union lines.[29] Over the fence the gray lines swept, now using the bayonet and beating back all in their path. The 130th Illinois, although almost destroyed, refused to quit and the fighting was now hand to hand.

Union troops began to turn and run, and finally the 130th Illinois Regiment, all but deserted by the rest of the Federals was completely destroyed.[30] These gallant men from Monroe, Sangamon, Alexander, Christian, Bond, Coles, Richland, Lawrence, Clark and Edgar counties in Illinois[31] were all killed, wounded, or captured and the regiment ceased to exist.

While this was going on, Terrell's regiment and the brigades of Major, Bagby, and Vincent were charging headlong into the right of the Union right flank, turning it back toward the center. Now General Taylor ordered Walker's Texas Infantry on the right side of the road to attack and Buchel's and Debray's cavalry regiments to push through to the Federal rear.[32] As Walker's men advanced, they fired point blank into the gunners and finished them off with the bayonet.[33]

Debray and Buchel were not able to make much headway in the dense forest, but Walker's brigades pressed on sweeping the field before them. They charged up the hill, driving back the 23rd Wisconsin and 67th Indiana and captured three guns of the Federal Nims' battery which were turned on the Federals.[34] The Union troops now were moving back into the woods, continuing to try to check the Confederates but without success.[35] At the edge of the woods about a half mile from the original line the Federals formed another line and received reinforcements when the third division of the 13th Corps arrived and were double-quicked into position.[36]

This second line held the Confederates for about an hour, but then with both flanks turned and the Confederates charging down the road in the center in two lines, supported by a line of cavalry, they broke, threw down their muskets, knapsacks, and other equipment and joined the remainder of the terror-struck first line in their panicky flight to the rear.[37]

The Federal cavalry wagons loaded with supplies for the army were only a short distance behind the second line on both

MANSFIELD BATTLEFIELD

sides of the narrow road, leaving barely enough room for artillery and troops to squeeze through.[38] But now they meant disaster for the Federals. The runaway men plunged headlong into the wagons. Teamsters attempted to turn the wagons, getting some stuck in a small creek. The road was blocked. No more artillery could be saved, no wagons could get by. It now became a complete rout.[39]

Terrell's regiment and the other cavalry on the left of the Confederate line had fought on foot and now it was necessary for them to return back behind the original line of battle to get their horses so they could chase the fleeing Federals.[40]

About three miles from the second battle line the Federals under General Emory with 5,100 fresh troops formed still a third line in an effort to stop the Confederates. These troops had not yet been in the battle. They were forced to fix bayonets to jab their way through the onrushing first and second lines of Union infantry and cavalrymen who were still fleeing through the woods by the hundreds, bareheaded, without equipment, some wounded and bleeding, but all trying desperately to get away from the Confederates.[41]

The Confederates immediately formed a battle line and again charged the Federals.[42] Green, Polignac, Major, Bagby, and Randal at the Federal right and Walker, Bee, and Scurry at their left. Again the Federals were driven back, but it was nearly dark now and this put an end to the fighting.[43]

Jaded after two hard days of fighting the weary men of Terrell's regiment made camp near the small creek which they had bought with blood a short time before. Forage for the horses was brought up from Mansfield and fires were lighted for cooking and reorganization of the companies of the regiment. When the bugler sounded assembly, some of the men of Terrell's regiment still lay on the original battlefield three miles to the north. The entire area was now one big camp of the various regiments. The roads were choked with wagons and the 1,423 Federal soldiers who were captured started northward. On every side of the road were the dead and the dying. The cries of the wounded could be heard long into the night. Some begging for a drink of water, some begging to be shot. The doctors and chaplains spent the

MANSFIELD BATTLEFIELD

night going over the battlefield with their lamps caring for the wounded, both Confederate and Federal.[44]

Not far to the south of Terrell's regiment and the other Confederates the Federals went into camp in line of battle. No fires were allowed and they had no blankets for the cool April night. They rested after the long day of battle and heard the Confederates calling roll; and later could hear the screams of the wounded from the battlefield not far away.

Terrell's Texas Cavalry Regiment had proven themselves in battle. They had faced the Federal muskets and cannons bravely and were now tired battle veterans. These men from East Texas were now an effective fighting regiment and the Federals would hear much more from them in the months to come.

General Richard Taylor reported the battle of Mansfield as follows:

> I soon found that the enemy was weakening his left and massing on his right to turn me. I at once brought Terrell's regiment of cavalry to the left to re-enforce Major, and Randal's brigade, of Walker's division, from the right to the left of the road to strengthen Mouton's, causing the whole line to gain ground to the left to meet the attack. These movements were masked by throwing forward skirmishers toward the enemy and deploying Debray's regiment of cavalry in the open fields on both sides of the road. It was not until 4 P.M. that these changes were completed, when, becoming impatient at the delay of the enemy in developing his attack, and suspecting that his arrangements were not complete, I ordered Mouton to open the attack from the left. The charge made by Mouton across the open was magnificent. With his little division, consisting of his own and Polignac's brigades, the field was crossed under a murderous fire of artillery and musketry, the wood was reached, and our little line sprang with a yell on the foe. In this charge General Mouton, commanding division, fell. Colonel Armant, of the Eighteenth Louisiana; Colonel Beard, of the Crescent Regiment; Lieutenant-Colonel Walker, commanding Twenty-eight Louisiana; Lieutenant-Colonel Noble, Seventeenth Texas; Major Canfield, of the Crescent Regiment, were killed,

and Lieutenant-Colonel Clack, Crescent Regiment, dangerously wounded. Seven standard-bearers fell one after another with the flag of the Crescent Regiment.

Despite these and other heavy losses of officers and men, the division never halted for a moment nor ever fell into confusion, but under the gallant Polignac pressed stubbornly on. Major, with his division, consisting of his brigade, under Colonel Lane, Bagby's brigade, Vincent's brigade of Louisiana cavalry, re-enforced by Terrell's regiment drawn from the right, dismounted his men on Mouton's left and kept pace with his advance, forcing back and turning the enemy's right. Randal supported Mouton's attack by advancing his regiments on echelon from the left. In vigor, energy, and daring Randal surpassed my expectations, high as they were of him and his fine brigade. These movements on the left of the road to Pleasant Hill were under the immediate direction of Maj. Gen. Thomas Green, who displayed the high qualities which have distinguished him on so many fields. As soon as the left attack was well developed I ordered Major-General Walker to move Waul's and Scurry's brigades into action, directing General Bee, on his right, to press on with Debray's and Buchel's cavalry to gain the enemy's rear. Believing my right outflanked by the enemy, General Walker was instructed to throw forward Scurry to turn his left and gain a position on the high road beyond his main line of battle. The dense wood through which Bee advanced prevented him from gaining much ground, but the gallantry and vigor with which that accomplished soldier (Walker) led his fine brigades into action and pressed on the foe has never been surpassed. Until he was disabled by a painful wound on the following day, every hour but illustrated his capacity for command. The enemy in vain formed new lines of battle on the wooded ridges, which are a feature of the country. Every line was swept away as soon as formed, and every gun taken as soon as put in position. For 5 miles the enemy was driven rapidly and steadily. Here the Thirteenth Corps gave way entirely and was replaced by the Nineteenth, hurriedly brought up to support the fight. The Nineteenth Corps, though fresh, shared the fate of the Thirteenth. Nothing could arrest the astonishing ardor and courage of our troops. Green, Polignac, Major, Bagby, and Randal on the left, Walker, Bee, Scurry, and Waul on the right swept all before them. Just as night was closing in the enemy massed heavily on a ridge overlooking a small creek. As the water was important to both parties, I ordered the enemy driven from it. The fighting was severe for a time, but Walker, Green, and other gallant leaders led on our tired men, and we camped on the creek as night fell, the enemy forced back some 400 yards beyond. The conduct of our troops was beyond all praise. There was no straggling, no plundering. The vast captured property was quietly taken to Mansfield and turned over untouched to the proper officers.[45]

General Bee reported his participation in the battle of Mansfield as follows:

> At 4:00 P.M. General J. G. Walker moved up his division of infantry, my cavalry covering his right, but being in the timber my progress was slow, and not until after the infantry had captured the wagon train and 21 pieces of artillery did I succeed in disentangling myself from the swamps and morasses. Then, moving rapidly, crossing the Seven-Mile Creek, following up what was reported to me as a routed foe, but at once encountered the Nineteenth Army Corps of the Federal army, who not having come up in time to take part in the action of Mansfield, was now fresh and well posted on the crest of the hill surmounting the creek. Owing to the road being blocked by the captured train, our troops of the different arms became intermingled, but imbued by the proper spirit, acting as one organization, infantry and dismounted cavalry charged the enemy's line and maintained their ground until night put an end to the most severe action of the day.[46]

Private H.C. Medford of Lane's Texas Cavalry, fighting dismounted on the left of the Confederate line reported the Mansfield action in his diary:

> At or a little after four o'clock the signal for a general engagement is given. Brigade after brigade of the infantry

is put in. The strife begins to be serious. Artillery handsomely playing on both sides. The elements begin to fry with the passing of balls . . . General Major gives his division orders to charge the enemy . . . The ball is now fully open, and all hands are engaged.

Our commanding officers are all upon the field and every man in his place. The strife of this battle is terrible. Many of our men are falling. The whole heavens are replete with destructive missiles. There is not a safe place anywhere upon this battlefield. The enemy stubborn — seem to move but little. At length they give way and our forces drive them back with great slaughter.[47]

A member of Walker's Texas Division reported the Mansfield battle as follows:

Shortly after the report of General Mouton's death, the cavalry mount, and move off to the right, in full gallop. Presently, General Walker and staff are in their saddles. He orders his brigade commanders to prepare for action. All being in readiness, he gives the command: 'By the right of companies to the front, forward march!' Every man moved off quickly, with a confident and determined step. The line of march was through a large field in our front, then through a skirt of timber, and into another field. Picture a nearly triangular space, broken by woods, fences, and fields, — its base a long ridge of underbrush running from southeast to northwest, its lower side traced by a line extending westerly to a line of woods that forms the left right-angle as you approach the area by a road, the Mansfield and Pleasant Hill highway, which intersects with that area. As we approached a narrow skirt of timber, and about six hundred yards from the enemy's position, we beheld General Walker, mounted on his iron-gray horse, with his field-glass to his eye, taking observations of the enemy's position. His actions and features were a study for the closest scrutinizer of physiognomy. Not a quiver on his face — not the movement of a muscle, to betray anxiety or emotion, notwithstanding the shower of balls whizzing around him.

Resting a few minutes in the skirt of timber, the command was given, 'By companies, into lines! ' After the line was formed, orders were given to 'fix bayonets.' In the meantime, the enemy continued firing upon us from their batteries. Soon the command was given to 'double-quick.' We immediately commenced advancing in the direction of the enemy, who were securely posted behind a rail-fence. They greeted our coming with a perfect shower of leaden hail. The men shouted, at the top of their voices, at each iron messenger as it approached, many indulging in jokes and witticisms, such as, 'This kind of ball-music is fine for dancing.' 'Here comes another iron pill.' 'Dodge, boys, but don't tremble!'

The fire of the enemy increases; it is terrible. He is gathering all his strength for one final struggle. Shells, canister, and bullets are falling around like a hail-storm. Our different brigade commanders rode along their lines, encouraging their men; still there is no faltering, but wild cheers, and on they press. When our army had arrived within about fifty paces, and before we had fired a shot, a general flash was seen along the enemy's line, and a storm of bullets went flying over our heads. They had aimed too high. Onward our troops advance, pale with excitement, compressed lips and blazing eyes beckoning the spirit of their determination. Casting your eyes along the column, you behold the flags of the various regiments floating on the breeze, and each regiment trying to be the first to scale the fence. Nearer our troops advance; the color-sergeants flaunt their flags at the enemy, and fall; others grasp them and fall, and they are then borne by the corporals. In this fearful charge, there was no flinching nor murmuring — nothing but the subdued talk of soldiers, the gritting of teeth for revenge, as they saw their comrades falling around them. At last the fence is gained; over it our troops go, like an avalanche of fire! A loud and prolonged Texas yell deafens the ear; their cheers rise in one great range of sound over the noise of battle, and are heard far down the lines to the left, where the Louisiana boys are at it.

Nothing could withstand the impetuosity of our charge. After crossing the fence, we came abruptly upon the enemy's guns. With loud huzzas we rushed upon the enemy before they could reload. A murderous discharge of rifle-balls was poured into their very bosoms; afterwards, using our bayonets, we mercifully bayoneted them, ere they could recover from their astonishment. Their prostrate column was trampled in the mire. Ah! now come the tug of war. The enemy is panic-stricken; they abandon their artillery; they cannot stand the bayonet charge; they retreat, and from their apperance, 'every man is for himself.' They sadly feel the loss of their artillery. Cheer after cheer bursts forth from our lines, as the enemy is seen fleeing, casting away their knapsacks and arms. Our cavalry now charges down on their flanks, making the very ground quake and the enemy tremble. Urged on by the excitement of victory, we pursue the flying foe, killing where they dare resist, and capturing them by hundreds. . . . After pursuing them four miles, they finally made a stand at a peach and plum orchard, where they were reinforced by the 19th Army Corps. Entirely unconscious of the arrival of fresh troops to their assistance, we passed half-way through the field before we became aware of their reinforcements. Then came the terrible shock. Volley after volley, and shower after shower of bullets came whizzing down upon us. It was utterly impossible to advance, and to retreat beneath the range of their long guns seemed equally desperate. We lay down, arose again and then involuntarily sought such shelter and protection as the ground afforded.

. . . . Our brave men attempted again and again to charge the enemy, who were behind their barricades of logs and fences, which they hastily constructed to cover their retreat; but human fortitude and human bravery were unequal to the task. The very air seemed dark and hot with balls; the thunders of the artillery guns resounded through the heavens and seemed to shake the earth to its very center, and on every side was heard their crushing sound as they struck that swaying mass, tearing through flesh, bone, and sinew. The position of our line could have been traced by our fallen dead. Within a few short moments many a gallant spirit went to its long home.[48]

In his book, *Campaigning With Banks in Louisiana, '63 and '64,* Federal soldier Frank M. Flinn described the battle as follows:

> . . . High and dreadful swelled the conflict. The enemy pressing forward at all points met a terrible resistance. Volley after volley was poured into their ranks, sweeping down hundreds, only to give place to new hundreds, who pressed forward to supply the place of the fallen.
>
> Our troops stood firm, but the rebels, who outnumbered us more than two to one, began, after an hour's hard fighting, slowly to gain ground, and our thinned and bleeding ranks were pressed back by overwhelming numbers into the woods.
>
> The rebels now began to show a heavy force on our left, which was the real point of attack, their movements toward our right having been a ruse to induce us to weaken our left by sending troops to the right, in which they had succeeded. It was plain to all that no human bravery or skill could long withstand the odds against which our troops were fighting, and that unless Franklin speedily arrived, we would be forced to retire. Gen. Franklin, with his staff, did come up, but his division, under command of Gen. Emory, was yet in the rear.
>
> Our thinned and wearied ranks stood up nobly against the masses and murderous fire of the rebels, and cheer after cheer went up, mingled with the almost incessant roll of musketry and roar of cannon. The forces of the brave Gen. Ransom had been cut up dreadfully, and he himself borne wounded and bleeding from the field; but still they held this position, fighting gallantly. Gen. Cameron's Division of the Thirteenth Army Corps arrived and hastened to the support of Col. Landruin's Division, but like bees from a hive the rebels swarmed upon it, and it was fast melting away under the storm of bullets that was continually rained upon them.
>
> Blucher at Waterloo was not more anxiously looked for than was Emory of Franklin's Corps upon that field. But he came not. We had now engaged less than eight

thousand men fighting a force of over twenty thousand men in their chosen positions. Emory was reported to be within two miles with his division, and rapidly coming up. The officers encouraged their men to hold the field until his arrival, and bravely indeed did they struggle against the masses that constantly pressed them upon both flanks and in front, but, borne down by numbers, their shattered ranks were pushed over the field and into the woods beyond.

The enemy had now driven back our left, and were within sixty yards of Nims' Battery, which was firing double charges of grape and canister, sweeping down the rebels in piles at every discharge. Gen. Lee, seeing that Nims' Battery, if it were not speedily removed, would be captured, by direction of Gen. Stone, ordered Col. Brisbin to have it taken from the field. The order came too late. Not horses enough were left alive to haul the pieces from the field. The cannoneers lay thick about the guns, and dead and wounded rebels in windrows before them. Two of the guns were dragged off by hand, and Lieut. Snow was shot down while spiking a third. Four of the guns of this battery could not be got off and fell into the hands of the enemy.

In the meantime our right was fiercely engaged, and our centre was being pressed back, and finally the right also gave way. Six guns of the Mercantile Battery, two guns of Rawle's G Battery, Fifty United States Artillery, two mountain howitzers of the Sixth Missouri Howitzer Battery, four guns of the First Indiana Battery, and six guns of Nims' Battery were left on the field.

Nims' Massachusetts Battery worked manfully. The veteran battery, the hero of seventeen engagements, always successful, but this time doomed to defeat, deserves to have its name written in letters of gold.

When the time was approaching that it could hold out no longer, each piece was loaded with a case of grape and canister, spherical case shell and a sack of bullets containing about three hundred. This hurled death and destruction into the ranks of the enemy, who wavered and fell back at every discharge of these fated guns. The battery lost twenty-one officers and privates, sixty-four horses and eighteen mules. Then came one of those unaccountable events that no genius or courage could control. Suddenly there was a rush, a shout, the crushing of trees, the breaking down of rails, the rush and scamper of men. Men found themselves swallowed up as it were in a hissing, seething, bubbling whirlpool of agitated men, who could not avoid the current. The line of battle had given way. Gen. Banks took off his hat and implored his men to remain. His staff officers did the same; but it was of no avail. Then the General drew his sabre and endeavored to rally his men, but they would not listen. Behind him the rebels were shouting and advancing. Their musket balls filled the air with that strange file-rasping sound that war has made so familiar to our fighting men. The teams were abandoned by the drivers; the traces cut and the animals ridden off by the frightened men. Bare-headed riders rode with agony in their faces, and for at least ten minutes it seemed as if all were going to destruction together. They rode nearly two miles in this madcap way, until on the edge of a ravine which might formerly have been a bayou, we found Emory's Division of the Nineteenth Army Corps, veterans who had never been defeated. The rock of safety to the Thirteenth Corps was drawn up in line of battle. Opening their ranks to permit the retreating forces to pass through, each regiment of this fine division closed up on the double quick, quietly awaited the approach of the rebels, and in less than five minutes on they came, screaming and firing as they advanced, but still in good order and with closed ranks.

All at once from that firm line of gallant soliders that now stood so bravely there came forth a course of reverberating thunder, that rolled from flank to flank in one continuous peal, sending a storm of leaden hail into the rebels' ranks that swept them back in dismay, and left the ground covered with their killed and wounded. In vain the rebels strove to rally against this terrific fire. At every effort they were repulsed, and after a short contest they fell back, evidently most terribly punished. It was now quite dark, and each party bivouacked on the field. Thus ended the battle of Sabine Cross Roads, April 8th, 1864.[49]

3

The Battles of Pleasant Hill and Blair's Landing

After the Mansfield battle ended General Banks[1] initially decided to retreat no further and gave orders for the Union forces to prepare to resume their northward drive toward Shreveport on the morning of April 9. However, at a meeting with his officers later in the evening, he was convinced by them that the army should retreat to Pleasant Hill where they could get water, and this is the course he followed. About midnight the army began their march toward Pleasant Hill and did not arrive until about 8:30 a.m.[2] The Confederates were surprised to find that the Federals had retired during the night.[3] So silently had it been done that the sleeping Rebels did not know when they had left the area.[4]

General Thomas J. Churchill's[5] Missouri infantry division, along with General Parsons'[6] Arkansas division, together about four thousand strong, from General Price's command in Arkansas, had been ordered to march for Mansfield at dawn on April 8 from Keatchie, twenty-two miles north of Mansfield and now reported to Taylor four miles north of Mansfield. They moved immediately toward Pleasant Hill at 3:00 a.m. with two days' rations.[7] The wearisome footsore men ambled on toward Pleasant Hill and at daylight many of the regiments were just passing through the battlefield of the previous day where the dead were still lying where they had fallen. Hundreds of stragglers and others were plundering the battlefield, stealing what they could

THE BATTLLE of Pleasant Hill, Louisiana, April 9, 1864. Sketch by C.E.H. Bonwill.

(From Leslie's Newspaper, 1864)

find of value from the pockets of the dead men and sometimes taking their clothing and footwear.[8] The fields were also littered with dead and wounded horses and the wounded men were still being cared for by the surgeons. Field hospitals had now been set up and the surgeons were busy piling up arms and legs as they worked urgently to save as many lives as possible. Bloody men were lying everywhere, ambulances were busy transporting wounded, both Confederate and Federal, toward the city of Mansfield where every house, church, and other suitable building was filled with the wounded. The road southward was almost impassable with wagons, troops, and prisoners.

In the meantime, Terrell's regiment and all of the cavalry were in pursuit of the fleeing Federal army, following them along the Pleasant Hill Road.[9] For twelve miles, the Confederates chased the Federals, capturing stragglers and seeing the evidence of the Federal rout. Arms and equipment were scattered everywhere and wagons which could not be saved had been put to the torch. Dead horses and broken wagons were everywhere and most of the homes along the road were destroyed, leaving the women and children standing by them not knowing what to do.[10] As the cavalry reached the open ground in front of Pleasant Hill the last of the Federals were entering the little village,[11] which was located on part of a plateau a mile wide from east to west. The Federals had taken advanced positions in a gully bordered by a thick forest of young pine trees located in front of some open ground several hundred yards wide. The main Federal line and cannons were back across the open ground on a plateau.[12]

Upon reaching the Federal lines Terrell's regiment and the rest of the cavalry dismounted and kept them busy while the Confederate infantry moved up.[13] Feints were made at the Federals at different points in order to determine the extent of their battle line. The Federal force consisted of General A.J. Smith's force of about 8,000 infantry, not engaged on the previous day, plus 6,000 of the Nineteenth Corps and 1,000 cavalry, totaling about 15,000.[14] The Confederate force was about 12,500, but the difference in physical conditions of the men in the armies cannot be overlooked. Taylor's men were completely exhausted by the fighting and marching of the two previous days. On the other hand, most of the Federals were fresh, not having been in battle on the day before, and those which had been in battle had a good rest while the Confederates were making the long march from the Mansfield battlefield of the previous day.

Four cannons of the 25th New York Light Artillery were across the Mansfield Road, with the 24th Missouri Infantry just behind them. The 14th, 27th, and 32nd Iowa regiments were placed in echelon to the west of the same road immediately behind the 24th Missouri. On the Federal right of the road were several New York regiments, the 116th, 114th, 161st, and the 153rd, plus the 29th Maine. Behind these troops on the right of the road were the 13th Maine, 47th Pennsylvania, 15th Maine, and 160th New York. Back to the west of the previously mentioned Iowa regiments were the 165th New York, 173rd New York, 162nd New York, and 30th Maine Infantry Regiments in that order along the battle line.

To the west and a hundred yards or so to the rear of these regiments behind an abatis were the 58th Illinois and the 119th Illinois and behind them about 300 yards was the 89th Indiana. General Banks' headquarters was about fifty yards to the right of the Indiana regiment. About 300 yards behind the 89th Indiana and General Banks' headquarters were the 49th Illinois, 33rd Missouri, 35th Iowa, and 178th New York covering the roads in the rear. On down the road, where it curved back toward Mansfield was the 47th Illinois, and to their right to the northeast about 500 yards down the Mansfield Road were the previously mentioned regiments of the 13th Maine, 47th Pennsylvania, 15th Maine, and 160th New York, which were being held in reserve. The 5th Minnesota and 8th Wisconsin were immediately behind the 47th Illinois, apparently being held in reserve, too.[15]

Terrell's Cavalry regiment along with the cavalry regiments of Hardeman[16] and McNeill[17] were on the far right of the Confederate line facing the 58th and 119th Illinois regiments initially, but they had orders to push down the Fort Jessup Road and once it was reached attack the Federal line of retreat.[18] This

COLONEL AUGUSTUS C. Buchel, mortally wounded at the battle of Pleasant Hill, Louisiana, April 9, 1864.

(Picture courtesy Robert W. Stephens, Dallas, Texas).

would have been a formidable job for them since they would have had to contend with no less than seven infantry regiments before reaching their objective, not to mention many Federal regiments to the rear who were not even on the battle line.

General Churchill's Missouri division and General Parsons' Arkansas division were to flank the Federals' left, driving to the Jesup Road, where they would then attack from the south and west. Walker's Texas division was to the left of Churchill and on the left of Walker was General Bee with Buchel's and Debray's cavalry regiments on the road, ready to charge through Pleasant Hill whenever the attack on the right had commenced. The cavalry of Major and Bagby (dismounted) under General Green was to the left of the Mansfield Road, with the objective of outflanking the Federal right and moving on to hold the Blair's Landing Road. Polignac's division was held in reserve on the road behind the cavalry of Buchel and Debray.[19]

All along the Confederate line there was a metallic rattle as bayonets were fixed on rifles. Ammunition could not be wasted. Terrell's regiment was on the extreme right end of the battle line dismounted along with Hardeman's and McNeill's regiments to Terrell's left and behind him, next to Churchill's infantry division. Horses, except for the officers and couriers, were left behind due to the dense pine forest. Churchill's division and Terrell's and the other cavalry moved southward at 3:00 p.m., reaching the Sabine Road after about two miles. Driving eastward toward the Federals they soon discovered that they had not moved far enough to the right in order to flank the Federal line and were forced to stop, move to the right again and then move again toward the Union line. The sound of their attack was to be a signal for Walker's division to move in echelon of brigades from the right. Mounted, Buchel's and Debray's cavalry regiments waited on the road ready to charge through Pleasant Hill.[20]

Artillery opened at about 4:30 p.m. as twelve guns sent solid shot and grapeshot exploding on the Federals. Federal guns on the Mansfield Road returned the fire with good effect,[21] but were soon overpowered by the Confederate guns and were forced to withdraw in panic, leaving the guns behind.[22] The

Confederate guns then concentrated on the hill beyond where the Union guns had been.[23]

Churchill's division attacked as planned, assaulting the Union line which had formed in a creek bottom. After firing at the Federals, they charged into the blue line of infantry with bayonet and using their guns as clubs engaged in hand-to-hand fighting. The 165th New York Regiment broke and ran back up the hill and were soon followed by the 162nd and 173rd New York, the 30th Maine being the last to retreat. The retreat continued back to the safety of the line formed by the 35th Iowa, 33rd Missouri, 49th Illinois, and 178th New York regiments which were waiting in a battle line along the Sabine Road. However, Churchill did not know that his rapid advance had left the 58th Illinois on their right flank hidden in a gully thickly covered with brush and trees. The swiftness of the advance had also left behind the brigade of Arkansas troops which covered their left flank. Churchill's division was, therefore, vulnerable to attack from both flanks. Terrell's regiment along with Hardeman's and McNeill's regiments, with orders to push on, had left a gap between Churchill's right flank and their own cavalry regiments. Unaware of the three Confederate cavalry regiments, the 58th Illinois came out of the gully and attacked Churchill's right flank. Waiting in ambush the three Confederate cavalry regiments poured a deadly fire into the flank of the 58th Illinois, checking their attack on Churchill's men.[24] Colonel Hardeman was in command of the three Confederate cavalry regiments and after having checked the Federals, ordered them to continue the advance on the right of the Confederate line, their objective to push on and attack the Federals from the rear. However, the 58th Illinois, although shocked and staggered by the volley in their flank, continued their attack on Churchill's right flank when Hardeman's premature order to continue was given. Other regiments of Lynch's brigade attacked Churchill's left flank when they saw the attack by the Illinois regiment. Churchill was, therefore, practically surrounded and forced back into a gully. Now came a furious charge by General A.J. Smith's battle-hardened men and McMillian's brigade, along with some of Benedict's men.[25]

Fighting men do not turn and run because they are cowards, only about one man in ten is in this category. Rather, they run because they feel that they are not as strong as their enemy.[26] This was true whether the forces involved in the running away were Federal or Confederate. At Mansfield it had been the Federal soldiers who broke and ran. Now it was the Confederates. Churchill's men, seeing the hopelessness of their situation, threw down their arms and ran to the rear. No one could stop them. Back where the cavalry had left their horses, they took them away from the guards and galloped off to safety.[27] Only darkness prevented the Federals from following up their victory on this side of the line.

In the meantime Terrell's regiment spearheaded the three cavalry regiments under Colonel Hardeman down a country road toward the Federal lines. Colonel Terrell led the three regiments on horseback, with the men in double time on foot. Captain Warren's company was the first company in the line and with Colonel Terrell in front were closest to the Federal lines. The tall pine forest was on each side of this little road and when the road turned abruptly to the left, Terrell and Company H were not visible to Colonel Hardeman following with the remainder of Terrell's regiment and the other two regiments. At about this time it became apparent to Colonel Hardeman that they were in danger of being surrounded and captured if they did not immedately move to the left and he so ordered this movement. Lieutenant Colonel Robertson of Terrell's regiment immediately executed the order with the remaining companies, but forgot to send word to Colonel Terrell and Company H, who were now out of sight. As a result Colonel Terrell and all of Company H were cut off by the Federals and spent the night making their way back around the Federals to the Confederate lines.[28]

But while the Confederate right side of the line was badly beaten, the left side held fast. General Bee waited on the Mansfield Road with the splendid cavalry regiments of Debray and Buchel. His signal for a cavalry charge through the village was to be the sound of Churchill's attack on the right. The rattle of musketry and Walker's advance from his position caused

General Green to think it was time for the charge.[29] He sent a courier to order Debray and Buchel to charge, but Colonel Buchel had not been sitting idle during the first part of the battle and had done some reconnoitering near the Federal lines. What he found was the Federal 24th Missouri infantry hidden in the gully behind a fence and he so informed General Green, telling Green that unless they could be routed, he was not certain they could be successful in their charge and more specifically that " . . . I can effect nothing but a sacrifice of men in a charge right now from my position." General Green sent the courier back to Buchel with orders to go ahead with the charge at once. Colonel Buchel said "Boys, we cannot disobey," and with Debray's regiment in the lead the two regiments galloped in column of fours up the road, intending to form a skirmish line and charge through the Union lines.[30] When Debray's regiment was directly in front of them, the Missourians blasted them from the side and other Federals blasted them from the front, catching them in a deadly cross fire of several hundred rifles. Debray's regiment was staggered. Scores of men were literally shot out of the saddle. Others went down with their horses killed or wounded under them. Colonel Buchel stopped his regiment in time to avoid the disaster, and dismounting them he attacked the Federals from the rear, forcing them back to their own lines. It was here, however, that Colonel Buchel was mortally wounded,[31] lying where he fell in the confusion of the battle as his regiment drove the 24th Missouri back to their lines. Later the same day the Federals counterattacked and retook the position where Buchel had fallen. They found him, but believed he was dead. General Bee assumed command of Buchel's regiment and led them again against the Federals, recapturing the position where Buchel lay wounded and unconscious.[32]

The fighting continued all along the line until nightfall, but slowly faded and finally died out. General Bee picketed the battlefield with companies of Debray's regiment. Mechling in his journal says that the pickets were two companies from Debray's regiment and two companies from Buchel's regiment.[33] An occasional shot by the pickets as late as ten o'clock was reported by General Bee.[34] Debray says shots were fired until daybreak.[35]

General Banks retreated toward Grand Ecore during the night. General Taylor withdrew all of his troops, except the four companies on picket, to the mill stream about six miles to the rear. Most of the cavalry was sent on to Mansfield where there was forage and water for their horses.[36] Terrell's regiment returned to Mansfield with most of the other cavalry and were joined there the following morning by Colonel Terrell and Captain Warren with Company H. The regiment had been in continuous movement for three days, the last two days of which they were directly engaged in battle. Men and horses needed a rest. Major Morgan had been seriously wounded in the arm during the battle of Pleasant Hill. General Taylor reported Colonel Terrell as being wounded in his report of the Pleasant Hill battle,[37] but Terrell in his "Chronology of the War"[38] says that this is incorrect. Apparently the reports received by Taylor had confused Major Morgan with Colonel Terrell. The regiment suffered four killed, fourteen wounded, and ten missing during the two days of battle April 8 at Mansfield and April 9 at Pleasant Hill. This was light when compared to some of the infantry regiments and to the two sister cavalry regiments of Buchel's First Texas Cavalry Brigade whose casualties were nine killed, fifty-one wounded, and four missing, or a total of sixty-four; and Debray's Twenty-sixth Texas Cavalry who had nine killed on the battlefields, sixty-three wounded, and seven missing, for a total of seventy-nine.[39]

Sextons were still burying the dead the following frosty morning at the Mansfield battlefield.[40] By now, two days after the battle, their pockets had been picked clean. Common graves had been prepared for the Federal dead and Confederate dead separately. Bayonets with the tips turned up were used to hook the collar of the dead men and drag them to their graves.[41] In Pleasant Hill some of the Federal and Confederate dead were being buried in a little cemetery near the stagecoach station. There also the surgeons were busy with their saws and knives where many of the wounded died on the operating tables.

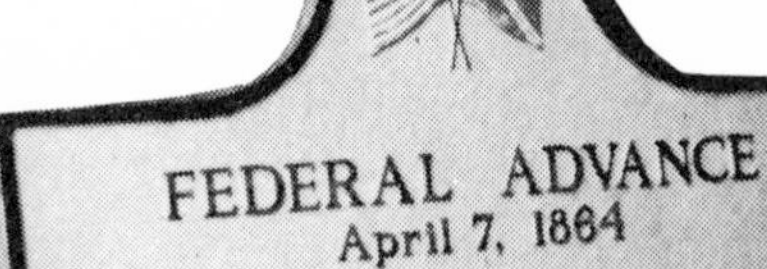
FEDERAL ADVANCE
April 7, 1864
Ten-Mile Bayou
Here the Confederate Cavalry commanded by Maj. Gen. Thomas Green stopped the Federal Advance in a stiff fight ending at dark on April 7. The two forces bivouacked on opposite sides of this stream. Green withdrew most of his forces before dawn to a point near Mansfield, leaving Gen. Bee with two regiments to delay the Federals. The 13th and 19th Corps moved from Pleasant Hill to this point on the morning of the 8th.
ROAD TO CEMETERY
This road leads to old cemetery where soldiers of both armies, who fell in the Battle of Pleasant Hill are buried.
CEMETERY
After the Battle of Pleasant Hill many brave men were put to rest here. Some wore gray, some wore blue.
THE OLD CISTERN
Both Gen. Taylor's and Gen. Banks' troops obtained drinking water from this cistern while each army occupied Pleasant Hill during the Red River Campaign in April 1864. The stage coach station stood a few feet north of here.

THE BATTLE OF BLAIR'S LANDING

While Terrell's regiment rested at Mansfield, General Bee with Buchel's and Debray's regiments were in pursuit of the Federals toward Grand Ecore. General Green was soon given orders by General Taylor to proceed to Blair's Landing about sixteen miles from Pleasant Hill to wait for the Federal fleet of about thirty gunboats and transports, which would be turning back from their original destination of Shreveport and heading back toward Alexandria now that the Red River expedition had failed.[42]

Woods' 36th Texas Cavalry and Gould's 23rd Texas Cavalry along with portions of Parsons'[43] cavalry brigade had only reached Mansfield on the evening of April 9 and morning of April 10 so had not as yet been engaged in the Red River fighting. Terrell's regiment which had returned to Mansfield for forage, and Wood's, Gould's, and the elements of Parsons' brigade were all rushed to General Green on April 10 and early April 11.[44] Some artillery was also sent to General Green, but there were no pontoons for crossing the Bayou Pierre, which was about four miles from Blair's Landing on the Red River where he was to engage the Federal fleet. General Green marched for Blair's Landing at 6:00 p.m. on April 11. Because he had no pontoon to get over the Bayou Pierre he was able to get only three of his guns across and part of his cavalry.[45] Terrell's regiment had been in both the battles of the previous two days and the other cavalry troops with them had not been in any of the conflict as yet, so Terrell's regiment was used as flankers on each side of the regiments of Woods, Gould, and Parsons, the only Confederate troops directly engaged at Blair's Landing.[46]

The Confederates reached Jordan's[47] on the Bayou Pierre at about one o'clock on April 12, dismounting and leaving every tenth man with the horses. They crossed the bayou at 3:00 p.m. in whatever boats they could find in the vicinity. As the last boat was about to leave, Colonel Parsons rode up, having made a full forty miles to be there in time for the battle. In a column the troops continued toward Blair's Landing until they halted near the open plantation and loaded and capped their guns. After a rest they continued through the timber, crossed on a bridge over a little bayou, and came to the edge of a clearing. The three guns were ordered to move out to the mouth of the lane about 400 yards from the river. General Green and Colonel Parsons trotted the main body of men toward the river in a column until they reached the artillery. Here they formed into a battle line.

Upon command of General Green the little force charged across the open ground toward the boats on the river. Grape and canister from the gunboats forced them to stop about twenty paces from the river but they continued to pour a hail of Minié balls into the Federal boats. Transports on the river raised the white flag several times, but the gunboats, since they were protected with iron plating, kept up the fight, forcing the Confederates to continue their attack.[48]

Everything that was made of wood on the gunboats *Osage* and *Black Hawk* was riddled with bullets. An iron shield in the pilot house of the *Black Hawk* was found later to have sixty bullet marks on it.[49]

For almost an hour the firing was continuous between the gunboats and the Confederates on the river bank. General Green had dismounted and was urging the men of Woods' regiment on when a shell from the *Black Hawk*, about forty yards away, burst near him, and an iron ball about the size of a big marble hit him in the top of the head, tearing away about half of his skull.[50] The same shell burst cut Lieutenant Colonel Benton's right arm in two and wounded or killed two or three others nearby. Colonel Benton's arm was later amputated near the shoulder.[51] Colonel Parsons took command of the Confederate force immediately.

About the time of General Green's death the Federals had landed more troops on the opposite side of the river and above the Confederate position. These troops began to pour a hail of bullets into the Confederates. The battle continued until dark with cannon shot plowing up the ground and shells bursting over the Confederate force. Colonel Parsons then retired the men from the river bank, while the Federal boats escaped down the river.[52]

HAMILTON P. BEE

After dark Colonel Terrell and Colonel Hardeman took the body of General Green back of the Confederate lines.[53] Contrary to other reports, the Confederates lost only forty-eight men from Woods' and Gould's regiments, of which twelve were killed, thirty-two wounded, and four were missing.[54] Parsons' brigade had only two killed and seven wounded.[55] Terrell's regiment, although present, or nearby protecting the flanks, suffered no casualties.

4

Skirmishes after Blair's Landing

After spending the night camped on the Pierre Bayou, Terrell's regiment joined General Bee and the other regiments of Bee's cavalry division near Grand Ecore on April 15.[1] Here the regiment was engaged in hot skirmishing with the Federals, who continued to retreat southward.[2] On April 17 there was more furious skirmishing with the Federals and on April 18 several prisoners were brought in by the pickets. Terrell's and Yager's (formerly Buchel's) regiments were on picket duty in advance of the Confederate lines on April 19 and were not relieved until midmorning on April 20, when General Bee's division then moved toward Monett's Ferry, below Cloutierville. The division camped that night several miles above Cloutierville on the Cane River, sending out scouts and pickets in all directions.[3] Now the Confederate cavalry was in a position to cut off the retreating Federals.

General Bee had been ordered to fall back before the retreating federals to Monett's Ferry, about forty miles below Natchitoches on the road to Alexandria, where the Federals were heading. On the afternoon of April 21 the Confederate force reached a point on the Cane River one mile below Cloutierville and about six miles north of Monett's Ferry and pitched camp. No trouble was expected from the Federals as they camped for the night, but about two o'clock in the morning on April 22

Federal cavalry drove in the Confederate pickets, causing the tired cavalrymen to form a line of battle. The wagon train was sent on to Monett's Ferry.[4]

MONETT'S FERRY

During this period of time the Union troops were in a race with the Confederates to get to Monett's Ferry. But as the Federals marched, they also burned — houses, fences, barns, chicken houses were all set afire; even the cabins of the Negroes were fired. At night the burning buildings lighted the way for the Union army's retreat.[5] General Bee's Confederates won the race to the ferry and set up a strong defensive position on April 22. Terrell's regiment was sent on to Beasley's Station about twenty miles away to guard the wagon train.[6]

At daylight on April 23, the 2,000 Confederates at Monett's Ferry found themselves facing about ten times that number of blue uniforms across the Cane River. The Federals promptly began movements to turn the Confederate left flank and make a feint at the Confederate right flank, with a large force ready to strike from the center. General Bee also had word that a large force was marching to strike them from the rear. He retreated to Beasley's Station where Terrell's regiment had been sent.

SKIRMISHES AFTER MONETT'S FERRY

Six days' rations were issued to Terrell's regiment and the remainder of Bee's cavalry command on April 24 and the wagons were sent to the rear. At four o'clock in the afternoon they moved out with two days' of cooked rations, stopping on the Monett Ferry Road to graze the horses. The next day orders were received to proceed to McNutt's Hill, a high bluff on the road between Monett's Ferry and Alexandria. Camp was made that night about six miles from McNutt's Hill and Woods' and Chessum's regiments were sent on to surprise and capture the Federal pickets there. However, the Federals had already passed by McNutt's Hill by the time the Confederate cavalry arrived on

April 26 and having no artillery, they could only skirmish lightly with the division of Federal cavalry which was acting as rear guard.[7]

From April 26 to May 1 Terrell's regiment remained at McNutt's Hill,[8] but on May 1 marched at sunrise and encamped that night at Mather's Bridge. On May 2 they continued to advance toward Alexandria, skirmishing sharply with the Federals. Apparently the Confederates pushed too hard for on May 3 while to the west of Alexandria, the Federals advanced toward the Confederates in force with infantry and cavalry and drove them back. Terrell's and Likens' regiments were dismounted and formed into a line of battle, with the remaining cavalry acting as support. The two regiments charged the Federals with loud Texas yells, driving them back in confusion.[9]

The standing order was to "attack every day and to annoy the enemy by every possible means."[10] Many did not believe the cost in dead and wounded was worth the small damage done to the Federals during these attacks. But attack they did, every day as ordered. General Bee says in his report that:

> In the operations on Bayou Boeuf, for seven days my division, much reduced by loss in battle, sickness, &c., maintained an unequal contest with the best troops of the enemy, and by a series of attacks and annoyances actually prevented them from sending troops to remove the blockade which General Major so gallantly maintained on the river below Alexandria. So impressed were they that my force was a large one that they feared to leave it in rear, even when the stake was the very subsistence of their army. I claim for my troops (Gould's, Woods', Terrell's, Likens', Yager's, Myers', and Vincent's Louisiana cavalry) the highest praise for their gallantry, patient endurance of fatigue, and neverfailing enthusiasm; and the fact that they held their position for so many days, fighting the enemy's cavalry and driving them back on their artillery and infantry supports every day, and only abandoned it when the Sixteenth Army Corps of Banks' army, 10,000 strong, marched down on them, stamps them soldiers worthy of the cause for which we fight.[11]

BATTLE OF LECOMPTE

Fighting was principally on the Bayou Robert Road between Chamber's plantation and Alexandria on May 3 and May 4, with a brisk fight on May 5 a few miles from Alexandria.[12] Terrell's regiment was engaged in all these encounters and was in the advance on May 6, with Vincent's Louisiana cavalry and the regiments of Woods and Debray, attacking the Federals along the Lamourie Bayou south of Alexandria near Lecompte on the Boeuf Bayou. The Federals were in force and meant business, driving the Confederates from their positions back to Lecompte, where they camped for the night. Terrell's regiment had been at the key position in the affair at the Lamourie bridge. The next morning on May 7 the Federals were on the south side of Lamourie Bayou and looking for a fight. Opening on the Confederates with solid shot and shell, they then sent forward skirmishers in a brisk fight.[13] Terrell's regiment advanced at a trot about eight hundred yards, then coming to the edge of a field they spotted a line of Federal cavalry and Colonel Terrell ordered a charge. Federal infantry, however, was hidden in a gulley and at this moment jumped up and poured deadly fire into the regiment. Shocked, the regiment fell back, rallied, and remounted. Thousands of blue uniformed Federals then advanced on the Confederates, but were sent reeling back with a furious fire from the gray lines. Darkness ended the fight and the Confederates camped on the battlefield.[14]

THE REGIMENT MOVES TO MARKSVILLE

May 8 found the regiments again pecking away at the Federals and on May 9 Bee's division was relieved from duty in that area and ordered to Marksville, Louisiana, near the Red River for the same kind of duty. Arriving at one o'clock in the morning on May 10, they received orders to attack gunboats and transports on the Red River at daylight. However, a heavy rain lasted until about ten o'clock and the artillery did not arrive in time for the attack.[15]

THE BATTLE OF MANSURA

The Federal retreat continued on May 13 as they began to leave Alexandria at 7 a.m., setting fire to the town before leaving. Terrell's regiment and the other Confederate cavalry continued to attack, forcing the Federals several times to bring forward troops to outflank the Rebels, causing them to withdraw. Several members of the regiment were killed or wounded. By May 15, Marksville was reached and the Federals spent the night there. On May 16, the Confederates formed a battle line near Mansura with their little force of about 3,500 men, challenging the Federal army of 18,000. Thirty-two Rebel cannons were aimed at the bluecoats, ready to fire.[16] The battle lines were formed on a wide, smooth prairie where the entire Union army could be seen by the Confederates, magnificient in its columns and formations, one behind the other, mile after mile.

Bagby's brigade, consisting of Terrell's, Yager's (formerly Buchel's), and Likens' regiments of Texas cavalry, was assigned to cover the Confederate artillery, massed in the center of the line. Colonel Terrell was placed in command of the entire brigade for this battle, and formed a line in front of the Confederate artillery. The Federals charged Terrell's line, which feigned retreat and fell back behind the artillery as previously planned. Once behind the artillery safely, the cannons opened up on the charging Yankees with canister. The Federals fell back. Forty-two Union cannons were then brought up and formed in a single line facing the Confederates, less than a thousand yards away. Thunder roared for four hours from the seventy-four Federal and Confederate cannons. Then, the Federals began a flanking movement on the left of the Confederate line which now began to fall back before the overwhelming numbers. "Old Gotch" Hardeman whose Confederate cavalry was on the right of the line assessed the situation and at once made a daring move, he galloped to the left between the Union and Confederate lines, enabling the artillery to fall back, as the forty-two Yankee cannons continued to fire at them. Hardeman's regiment then formed a battle line and held the Federals back until all the Confederates were safely away. They then fell back coolly, leaving the field to the Federals. Although this had been a spectacular four-hour show, little was accomplished by the Confederates, and the Federals continued their retreat toward Simmesport.[17] Several members of the regiment were wounded and there was at least one man killed in this action.

THE BATTLE OF YELLOW BAYOU

General John Wharton, who was later to be shot by Colonel George W. Baylor in the Fannin Hotel at Houston, succeeded General Green as head of all the Confederate cavalry and continued to press the retreating Federal army. The Federals were crossing their artillery and wagons over the Atchafalaya Bayou on a bridge of boats and ferrying the troops in other boats. They would soon be safe from the Confederates. But General Wharton determined on one last battle before the Federals got away. Outnumbered five to one, the Confederates formed a line of battle and attacked the Federals, driving in their skirmishers and pickets. Terrell's regiment was held in reserve to be rushed into the first break in the Confederate lines. This occurred on the right, where Colonel G. W. Baylor's cavalry was almost surrounded by the Union troops. Terrell's regiment came up in time to enable Baylor to get out with only twenty-eight captured and a few killed, although he lost heavily in wounded. Overwhelming masses forced the Confederates to retreat and let the Union army continue their retreat.[18] The affair was costly to the Confederates. They lost about 450 in killed and wounded and 160 in captured. Federal losses were about 350 in killed, wounded, and missing.[19]

The last Federals crossed the Atchafalaya on May 20, and now the Confederates could not annoy them any more. At last the Red River Campaign was over.

DUTY AFTER YELLOW BAYOU

Terrell's, Yager's, and Likens' cavalry regiments moved to near Fort DeRussy after the battle at Yellow Bayou. Here they kept the Federals under surveillance and prevented any attempt by them to move back up the Red River.[20] Picket duty along the Atchafalaya occupied much of the regiment's time now, the men occasionally trading shots with the Federals across the river. There was also time to rest and write letters home. For those not on picket duty, drill was the order of the day. Several moves were made during the month or so after Yellow Bayou.

SKIRMISHES AT MORGAN'S FERRY AND CAPTURE OF SEVEN MEN

On August 25, Lieutenant Colonel Gurney of the 2nd New York Cavalry, with a party of fifty cavalrymen made a reconnaissance to Morgan's Ferry on the Atchafalaya River, about twenty miles south of Simmesport, Louisiana. Four Confederate cannons were in position on the Rebel side of the river, with a small protecting force of men. Two miles on the Yankee side of the river Colonel Gurney's New York cavalry encountered a Rebel picket consisting of a lieutenant and six men who they chased back to the river, where they took cover under their guns on the opposite side of the bank. These were all men of Terrell's regiment.[21]

Gabriel D. Gilley of Company H was the lieutenant, and the six enlisted men were: Privates Ballard A. Day, James L. Felton, and William Zuber of Company H; Privates John T. Rice of Company E, Corporal George Reager and Private K.H. Barbee of Company I.[22]

Colonel Gurney charged up to the river with half of his cavalrymen in the face of the artillery on the other side of the river and captured the seven Confederates.[23]

Lieutenant Gilley and the six enlisted men were sent back to start their long journey to Federal prisons on August 27. Gilley wound up in Boston Harbor on December 21, 1864. He survived and was paroled on March 13, 1865. Day, Felton, Zuber, Rice, and Reager all were sent to the Federal prison at Elmira, New York, and Barbee wound up in a hospital in Richmond, Virginia, where he was furloughed
1865. Day was released on July 7, 1865, Rice on June 19, 1865.

Felton died on April 20, 1865, of pneumonia and was buried at Elmira. Zuber died of diarrhea February 26, 1865, at Elmira, and Reager died at the same place in November, 1864, of the same ailment.[24]

COLONEL TERRELL GIVEN BRIGADE

Changes in command in the Trans-Mississippi department were made in August, 1864. General S.B. Buckner became commander of the Department of West Louisiana, General Walker was returned to Texas, and General Magruder was transferred to Arkansas. Early in September Colonel Terrell was made commander of a brigade of Texas Cavalry troops under acting Brigadier General Bagby, who was promoted to division commander. Terrell's brigade consisted of his own regiment, the 34th Texas Cavalry,[25] Yager's 1st Texas Cavalry, and Likens' 35th Texas Cavalry.[26] The brigade was on the line of the Atchafalaya River somewhere between Simmesport and Morgan's Ferry, about a twenty mile front. Yager's regiment was the regiment formerly commanded by Colonel Buchel, fatally wounded in the Pleasant Hill battle. This regiment was one of the few well-disciplined, crack cavalry regiments operating west of the Mississippi; so it was an acknowledgment of Colonel Terrell's cavalry know-how for him to be named commander of this brigade and an acknowledgment of the effectiveness of Terrell's regiment to be brigaded with such a regiment.

FIGHT AT MORGAN'S FERRY

On September 17, 1864, Terrell personally commanded a strategic position with a small part of his brigade and one battery of artillery at Morgan's Ferry on the Bayou Atchafalaya, about six or seven miles west of Morganza, Louisiana, and twenty miles south of Simmesport. Nims' battery and about 4,000 Federals attacked them furiously at about 2:00 p.m. but were unable to dislodge them. The Atchafalaya and the crack Texas sharpshooters prevented the Federals from getting across.[27] After the Federals withdrew, Terrell had his artillery moved to a higher point until he could construct breastworks for protection.

Orders were received by Terrell the following day from the commanding general of the cavalry south of the Red River to "risk nothing in the way of being captured, and that if an attempt is made to cross in above or below to fall back at once."

Expecting another attempt to cross by the Federals, Terrell called the remainder of his brigade to concentrate at Morgan's Ferry. This gave him about 1,200 men to face a Federal force of about 4,000 men and twelve cannons. Terrell's biggest advantage was the Atchafalaya River, which could not be crossed by cannon or wagons without a bridge.

At 2:30 a.m. on September 20, the advance of the Federal cavalry and four of the cannons passed a secret post that had been established by Terrell, and which sent word to him that a large force was on the way. The apparent objective of the Federals was to drive Terrell's brigade away from the Atchafalaya. Terrell immediately moved his wagons and sick and wounded to the rear and got ready for a fight.

Daylight found the Union troops just out of range of Terrell's guns, building a bridge across Muscle Shoal Bayou to the south of Terrell's position, and also opening a road leading to the river below the mouth of Muscle Shoal Bayou. Two cannons had already been crossed over the river to the north of the Confederates, and with wheels muffled, had been moved to a position from which they could enfilade the Confederate's position. The Federal plan of attack appeared to be to outflank Terrell on his right and take him in the rear, or else to cut off his retreat when the attack got started.

Terrell watched the Federals for several hours unable to do anything against such overwhelming numbers and with his ammunition for his artillery nearly exhausted from the fight on September 17. As the Federal plan developed, Terrell ordered his artillery to retreat, and as soon as he saw the road to his south would soon be crossed by the Federals, he ordered his regiments back. Three companies remained behind to skirmish with the Federals through the swamp. By 3:00 p.m. the Federals succeeded in crossing the river about two miles below Morgan's Ferry and moved onto the position the Confederates had vacated.

Under such a large force of Union troops the Confederates, under written orders not to be captured, had no choice but to retreat. Terrell's 34th Texas Cavalry Regiment retreated toward Evergreen by way of the Atchafalaya and Washington Road and Yager's cavalry galloped on to Simmesport and fell back from that point toward Evergreen, leaving pickets on the road. Likens' regiment retreated by Faulkner's Ferry Road, destroying a pontoon bridge on the Bayou Rouge. Terrell did not take a stand in the swamp because he would be easily flanked by the Yankees, and, also having sent Yager's Regiment to the north, he could not communicate with them to get them to come into action if he had made a stand.

As soon as they were safe from capture, Terrell had the three regiments placed in line, one in rear of another one-half mile, and remained in this position until he was relieved by Brigadier General Debray. Terrell's regiment and the rest of the brigade then moved on to Evergreen.[28]

The Federals expected Colonel Terrell to make some sort of a demonstration at Morgan's or Logan's Ferry to support a crossing of other Confederates in October, but the Confederates knew that the Federals had captured a letter showing the points at which such crossing would be made, so they were not attempted.[29]

On Colonel Terrell's retreat from the Atchafalaya they destroyed one million dollars worth of Federal property, including the pontoon bridge destroyed by Likens' regiment.[30]

FIGHT ON THE ATCHAFALAYA ON OCTOBER 17

Terrell's regiment and the other two regiments in his brigade continued to patrol the Atchafalaya, and on October 17 another fight took place. Terrell apparently had spies working for him because he was always well informed on where the Yankees would attempt to cross. When they attempted to cross, Terrell was waiting for them in ambush, and a smart fight took place. The Federals mentioned this in a report from 2nd Lt. A. M. Jackson of the U.S. Signal Corps to the assistant adjutant of The Military Division of West Mississippi in New Orleans on

November 21, 1864. He states in this report that Colonel Terrell's spy in this case was a man by the name of Johnson.[31]

DUTY ON THE BAYOU BOEUF

Moving back to the Bayou Boeuf, Terrell's regiment made its headquarters there, while continuing to harass the Federals along the Atchafalaya. While at this point an inspection was made by the inspector general of the Trans-Mississippi Department. The report of this inspection is critical of Terrell's men because of the way they roamed over the country taking what they wanted.[32]

WINTER AT ALEXANDRIA

Raiding parties by the Federals diminished greatly so that Terrell's brigade was no longer needed on the Boeuf. Instead they were moved to the vicinity of Alexandria where they would remain for the winter, with patrols being sent out to combat jayhawkers, who were more of a terror than the Yankees. Many of the brigade attended a big dinner given by the ladies around Alexandria for the soldiers on Christmas Eve, 1864. The tables were loaded with many good things to eat such as turkeys, pigs, ducks, chickens, cakes, candies of all kinds, puddings, and pies.[33] This must have been a big boost to their morale.

In January the Federals began to concentrate troops in New Orleans and General Smith thought that they were going to move against Mobile, Galveston, or possibly up the Red River again. To counter these last two possibilities he moved Forney's division from Minden, Louisiana, to East Texas, where it could go either to Houston or Natchitoches, Louisiana, as circumstances demanded. Churchill's division was moved to Minden for the winter, and General Buckner's command, including Terrell's brigade, was headquartered at Alexandria with Louisiana, Missouri, Arkansas, and Texas cavalry in winter quarters.[34]

PREPARATION FOR THE FEDERAL SPRING OFFENSIVE

In February the brigade moved to a point about twenty miles above Alexandria.[35] The spring offensive was expected by the Federals and preparations got under way for defenses along Red River. All available cavalry was to be thrown in front of the Federals to delay their march as long as possible.[36] It was felt that the guns at Alexandria would not be effective against the Federal's ironclad boats, and that all the heavist guns should be concentrated at Grand Ecore, where the Confederates would also be able to use land troops. Such use of land troops would be impossible at Alexandria.[37]

LIKENS' REGIMENT TRANSFERRED TO TEXAS

General Smith determined in January, 1865, to dismount nine of the cavalry regiments of General John A. Wharton, now commanding the Texas cavalry, leaving Wharton with two divisions, one of which would be regular cavalry, and one mounted infantry.[38] Texas cavalry, except for Bagby's division in Louisiana (which included Terrell's brigade), was now all in Texas under General Wharton. Likens' regiment was a part of Terrell's brigade and it had been selected as one of the regiments to be dismounted since it was one of the most recently organized regiments — the criteria for dismounting, except in cases where superior qualifications in discipline, conduct, and efficiency as cavalry was present in junior regiments.[39]

Likens' regiment started for Hempstead, Texas, on February 19, 1865, to report to General Wharton.[40] At this time the brigade was located near Logansport, Louisiana, and Terrell's and Yager's regiments moved the next day back to Grand Ecore. The 7th Texas Cavalry regiment was transferred to Terrell's brigade, replacing Likens' regiment, to maintain its cavalry strength.[41]

FURLOUGHS

On about April 1, all of the men of the brigade who were without horses were given sixty-day furloughs to return home and remount themselves. Many of the men who did have horses were able to make trades with the men without horses whereby they gave the unmounted men their horses, and probably other consideration, to be able to go home.[42] Most of these men would never return because General Lee would surrender on April 9 and word of this would reach Texas on April 19. General Smith announced it to his command on April 21 and at the same time called on his troops to continue the war.[43]

5

Return To Texas

Sometime between April 21 and May 1, Terrell's regiment left Grand Ecore where they had been camped and headed for Texas. Up through the Mansfield and Pleasant Hill battlefields of a year before, on to Marshall through Shreveport they marched with their worn horses. May 12 found them camped near Marshall and the next day they were started on toward Austin by Colonel Terrell while he remained behind. Although desertions were taking place by the hundreds in all commands, Terrell's regiment remained loyal and intact and was still an effective mounted cavalry unit, ready and willing to serve Texas wherever they might be asked to go.[1]

On May 13 the governors of Louisiana and Arkansas met in Marshall along with Texas' governor Murrah's representative, Guy M. Bryan. General E. Kirby Smith had requested the meeting for the purpose of determining what course he should take now that Lee had surrendered. The governors advised General Smith to disband the armies and send the men home immediately.[2]

In the meantime, on the same day as the governor's conference, in South Texas near Palmetto, Colonel Rip Ford was leading a few hundred Texans in a cavalry charge on Federal troops in what was to be the last battle of the war. Ironically, Confederates would win this last battle as the Yankees broke and ran in a near rout, with the Rebels in hot pursuit, driving them all the way back to the Boca Chica. The firing died down at about dusk and one last Federal shell exploded near the Confederates, causing a young Confederate to fire his rifle wildly at the Yankees in what was the last shot of the war.[3]

After the governors' consultation with General Smith ended, Colonel Terrell and several other high ranking officers held a military conference to discuss the situation and decide what course of action should be followed. General Smith was not present at this meeting and had no part in it. These officers felt that it would be useless to try to keep the various commands together any longer, but that each officer would ask for volunteers out of his command to cross the Rio Grande, join with Juarez in Mexico, and conquer the country up to the Sierra Madre mountains. They would then form their own government. It was felt by these officers that a small force of as few as 2,500 men could accomplish this and the divisions of Churchill and Price were still intact in South Arkansas and North Texas, and General Joe Shelby's command was camped at Marshall, ready to move. They also had twelve cannons currently moving westward from Marshall under Major Squires, and General Shelby also had a small battery of guns which would go with them. There were plenty of rifles and ammunition and stores of bacon and flour located at Corsicana, Austin, and San Antonio.

The group agreed that General Simon Bolivar Buckner would be asked to command them and made this offer to him, but General Buckner would not give them an immediate answer without the approval of General Smith. Smith at this time did not know that President Davis had already been captured, and felt that he must try to keep the commands together as he was sure President Davis would try to reach the Trans-Mississippi. General Buckner returned to Shreveport with General Smith without giving them a decision. This delay was fatal to their cause. Within forty-eight hours there would be few troops left from which to ask for volunteers and the dream of taking a slice of Mexico on the Rio Grande border would be ended.[4]

WILDCAT FERRY as it appeared in 1880.
(Picture courtesy Pioneer Village, Corsicana, Texas).

THE WILD CAT Ferry Crossing as it appears today.
(Picture by the author)

DISBANDMENT AT WILD CAT BLUFF

Terrell's regiment had marched on westward through Athens and stopped at Wild Cat Bluff, a small community on the Trinity River near where the Richland Creek runs into the Trinity.[5] Here on a high bluff overlooking the ferry on the Trinity River the regiment made its last camp. The news of the decision of the governors apparently reached them in record time, and on May 14, 1865, a little over two years after its formation, Lieutenant Colonel Robertson disbanded the regiment and told the men to go home. Taking their horses and arms, the men rode off by companies toward their respective homes. When Colonel Terrell arrived at Wild Cat Bluff on May 15 where the regiment had been camped, there was nothing left but one teamster with his headquarters wagon, its mules, and empty cartridge boxes.[6] Colonel Terrell was not even given the opportunity to bid his men goodbye.

BACK TO SHREVEPORT

His cavalry command gone, Terrell had no where else to turn now except back to Shreveport to the headquarters of the Trans-Mississippi Department. With his one still loyal teamster driving the headquarters wagon he raced back to Shreveport to report to General Smith, who was still there at this time. Rumors were now running rampant through the remaining Confederate commands. There was confusion and apprehension and the soldiers did not know what had happened or what was going to happen. The quartermasters began to issue large quantities of food, equipment, and rifles and ammunition and many of the soldiers slipped quietly from their regiments and went home.[7]

TERRELL PROMOTED TO BRIGADIER GENERAL

General Smith had returned to Shreveport after the governors' conference in Marshall, and on May 16 issued General Order No. 46 promoting Terrell to brigadier general.[8] Terrell accepted the promotion, not sure at this point exactly what the situation in the Confederacy was, and whether he would be given a new command. Now a general without troops, he and many of the other high ranking officers discussed the situation and the various alternatives open to them. They agreed to meet in Austin later and determine at that time what to do. There were rumors that all high ranking officers would be imprisoned by the Federals and in Austin they would better be able to assess the situation and make a decision. At this point there was also some chance that General Smith would continue the war till the end. General Smith also felt that President Jefferson Davis was on his way to the Trans-Mississippi Department in view of the fall of the Confederacy in the East, and unaware that Davis had already been captured by the Federals.

On the same day of Terrell's promotion General Smith received a wire that the troops at Galveston had mutinied,[9] so on the next day he decided to move his headquarters to Houston. By the time he reached that city on May 27, the Confederate army in Texas had completely disbanded and General Smith was also without a command.[10]

TERRELL GOES ON TO AUSTIN

After his promotion, Terrell rode his brown gelding[11] back through Athens to the ferry crossing at Wild Cat Bluff and on to Austin where he met General W. P. Hardeman, Colonel George Flournoy, Colonel Peter Smith, Colonel M. T. Johnson, General King, Major Hill (Terrell's former brigade commissary officer), and a Captain Roberts. These men and several other Confederate officers all decided to go to Mexico because they felt that an era of oppression was before the people, which they were powerless to do anything about. Governor Murrah shared their feelings and decided to go with them, although he was in the last stages of consumption. He was to meet them at their first camping place twelve miles south of Austin near the Tom McKinney crossing of Onion Creek, but he made it only as far as the house of McKinney as the ride had exhausted him and he was forced to stop. He eventually did reach Monterrey, Mexico, but died a few days after his arrival.

Governor Murrah gave Terrell a letter of recommendation to Maximilian, which was probably the last letter ever written by the governor. It was signed by Murrah and had the Great Seal of State attached. The letter was as follows:

Austin, June 15, 1865

''To His Imperial Majesty,
Maximilian, Emperor of Mexico.''

Sire:

This letter will accredit to your Majesty Generals A.W. Terrell and William P. Hardeman. The prolonged struggle made by the people of the Southern States of the American Union to preserve the right of local self-government has ended in their defeat. The above named gentlemen will seek through your Majesty's Empire for a soil and climate adapted to the agricultural products to which our people are accustomed. Should they return with a favorable report, a large number of men and their families would emigrate with them to Mexico. A victorious enemy is already approaching the Capital of Texas from the seacoast, and Generals Hardeman and Terrell will be but poorly equipped for so important a mission. Should they fail to find soil and climate they desire, I respectfully ask that your Majesty will facilitate their departure to Brazil.

In evidence of the authenticity of this, my official letter, I hereunto set my hand, and have caused the Great Seal of State to be affixed.''

Pendleton Murrah
Governor of Texas

(seal)

This letter was placed in an official envelope, wrapped in silk oilcloth and placed in the lining of Terrell's boot, where it was sewed up. Terrell and the other Confederate officers were then off to voluntary exile in Mexico.[12]

EX-CONFEDERATE GENERALS in Mexico after Civil War: Generals C. Wilcox, J.B. Magruder, Sterling Price, A.W. Terrell, and T.C. Hindman. A notation on this picture in the files in Austin indicates that Terrell weighed 140 pounds in 1865.

(Picture courtesy the Eugene C. Barker Texas History Center, The University of Texas at Austin)

TERRELL AND OTHER OFFICERS GO TO MEXICO

Each of the men in Terrell's party was mounted on a fine horse and most of them had an extra. They had good mules to carry their provisions and each was armed with a rifle and two revolvers. They had all agreed that they would not be taken by the Federals or be robbed by the predatory bands then prevalent over the state.

Early one morning they received word from two travelers from Roma that a part of Colonel E.J. Davis' Union regiment was at Roma with orders to capture or kill any Confederates who attempted to enter Mexico. The band of ex-Confederates decided to make a race for the ferryboat at the Alamo Crossing of the Rio Grande eight miles above Roma, ninety miles away at this point. By riding hard for thirty minutes, removing saddles and blankets and allowing the horses to graze or rest for five minutes, they made the trip to the Rio Grande by 2:00 a.m. the same night and crossed by swimming the river.[13]

In Mier, Mexico, Terrell had to give up his splendid brown gelding as a present to Liberal Force Colonel Garcia, commandant of the plaza, to enable the party to proceed on to Monterrey. A revolving rifle belonging to General King was also fancied by Colonel Garcia and this, too was given to him by Terrell without King's knowledge.[14]

Later, Terrell offered his pair of Colt revolvers to Colonel Garcia's son, who was their guide, as a reward for getting them safely to Monterrey, but young Garcia would not accept them.[15]

TERRELL IN MONTERREY

Arriving in Monterrey on June 28, Terrell's party stopped at the Paschal Hotel. The next day after their arrival, General J.B. Magruder, Colonel E.P. Turner, Major Oscar Watkins, Commodore Leon Smith, and Colonel A.C. Jones arrived, dusty and travel worn, and on this same day a banquet was given by the French officer in command, General Jeanningros at the governor's house and Generals Bee, King, Magruder, Hindman, Cadmus Wilcox, Walker, and Terrell, and Colonels Flournoy, Duff, and several others were invited to attend.

A few days later another party arrived with General E. Kirby Smith, General Preston of Kentucky, Governor Moore of Louisiana, General John B. Clark, Confederate senator from Missouri, Governor Reynolds of Missouri, Governor Allen of Louisiana, Confederate Congressman Parson, and Colonel Duff of Texas. General Sterling Price had not yet reached Monterrey.[16]

TO MEXICO CITY

Most of these Confederates sold their horses, many giving them up painfully as they had carried their masters through many fierce conflicts. They began to make their way individually and in groups to their own individual interest, many heading for Mexico City. Terrell's horse brought him $70.00, making him a total of $150.00, which he carried in a money belt. This money, his silver-mounted saddle, his saddle bags, with a change of linen, and a broad ornate sombrero given to him by the Mexican official in Mier to whom Terrell had presented his other horse, was his entire fortune. He boarded the stagecoach and headed for Mexico City.[17]

Terrell and other Confederates poured into Mexico City, most of them staying at the Hotel Iturbide, where General Magruder had his rooms. Terrell met Magruder here and showed him the letter given to him by Governor Murrah. An audience was arranged for him with the emperor's political secretary and General Magruder suggested to Terrell that he should procure suitable dress before seeing him. There for the first time since he became a Confederate officer, Terrell laid aside his faded gray uniform, spots showing where the gold lace had once been on the arms and under the stars on the collar, and put on a black frock coat and pants and a black silk vest, with patent leather cavalry boots and black silk hat. General Preston gave him a pair of lilac-colored kid gloves and General Magruder gave him a small cane with a woman's leg cut in ivory for its handle. His beard of four years was shaved, except for a long black moustache.[18]

GEN. TERRELL in later years.

MAXIMILIAN

Terrell's request for an audience with Maximilian was granted and he, along with Magruder and Maury, went to Chapultepec to see him. Terrell expressed his desire to settle in Mexico and that his associate, General Hardeman, was riding to Mexico City on horseback so that he could examine the country.

Later, Terrell was offered the provisional rank in the French army of *chef d' bataillon*, a rank with pay slightly higher than an American colonel. Terrell was to choose one other officer for the same rank and he chose Colonel George Flournoy. They were assigned to the staff of General Douay at San Luis Potosí to study the French language and French tactics. It was the feeling that war with the United States was inevitable, and Terrell and Flournoy would eventually organize their commands with ex-Confederates.

General Hardeman soon arrived and Terrell urged him to seek a position as surveyor with General Magruder. Hardeman left his fine thoroughbred chestnut stud horse with Terrell, who still had his silver-mounted saddle and was now mounted better than most men, even in the French army.

Terrell at this time was still unmarried, his wife, Ann, having died in 1860, and he was calling on a pretty young Mexican lady by the name of Gonzales in San Luis Potosí. Her brother later commanded the detail of soldiers whose duty it was to execute Maximilian. Later in 1871, Terrell would ride with this man in the same stage from Austin to San Antonio.[19]

DECISION TO RETURN TO TEXAS

Late in November, 1865, General Douay advised Terrell and Flournoy that an unpublished arrangement between Napoleon and President Andrew Johnson had been agreed to whereby the French would concentrate all troops at Vera Cruz and withdraw from Mexico in twelve months, and during that time the United States would not enforce the Monroe Doctrine. Terrell and Flournoy had their choice of being left in Mexico or going out with the French.

Terrell at once decided to leave, Flournoy decided to re-

GRAVE OF Gen. Alexander Watkins Terrell in Austin.

main. Terrell's decision to leave was influenced first of all by the secret agreement between Napoleon and Johnson; secondly, by a letter he had received from Colonel Thomas F. McKinney of Texas, who assured him that President Johnson wanted things returned to normal and that Terrell's friends wanted him to return and help in the reconstruction of Texas; and lastly by the "Black Decree" issued by Maximilian in October, 1865, which required that all republican officers captured by the imperial troops should be treated as bandits and shot. Terrell and the other ex-Confederate officers did not condone this sort of treat-

GEN. BUCHEL'S grave in Austin.

ment of prisoners, and this probably more than anything convinced him to return to Texas. Terrell also had young children in Texas, and admitted to the French that he was no soldier of fortune and wanted to return home rather than to go with them when they left Mexico.[20]

Travelling as "Colonel Monroe" with his sombrero and again wearing his faded gray Confederate uniform, Terrell returned to Mexico City, where he remained about ten days and then journeyed on to Vera Cruz, where he took a boat to Havana and New York.[21] Terrell must have arrived in New York in late November or early December, 1865, as he dined with General S.B. Buckner in New Orleans in December, 1865.[22]

FOOTNOTES

Footnotes for Chapter One

[1]Alwyn Barr, ed., "William T. Mechling's Journal of the Red River Campaign, April 7-May 10, 1864," *Texana*, I, No. 4, Fall, 1963, 365, 366; Rebecca W. Smith and Marion Mullins, editors, "The Diary of H.C. Medford, Confederate Soldier, 1864." *Southwestern Historical Quarterly*, XXXIV, Jan., 1931, 213.

[2]Commander of the Confederate District of Western Louisiana. Richard Taylor was born in Jefferson County, Kentucky, January 27, 1826. His early education was in frontier schools where he lived with his father, General Zachary Taylor. At the age of fourteen he was sent east to Massachusetts to complete his education and was graduated from Yale College August 21, 1845. He began his career as a planter in Mississippi in 1848, but floods ruined him and he moved to Louisiana, where he purchased a large sugar plantation in Saint Charles Parish. When the war began Taylor was commissioned colonel of the Ninth Louisiana Regiment, and hurried with it to Virginia, arriving a few hours too late to fight in the Battle of First Manassas. In October, 1861, he was promoted to brigadier general just before the Seven Days Battles, and was stricken with a partial paralysis. After a rest he was transferred to the Trans-Mississippi Department, where he exhibited a mastery of military science in the campaigns against General Nathaniel P. Banks. After the Red River Campaign ended, Taylor was transferred east of the Mississippi to command the Department of Alabama, Mississippi, and East Louisiana. He fought out the last days of the war with honor and finally surrendered the last organized body of Confederate forces east of the Mississippi to General E.R.S. Canby at Citronelle, Alabama, on May 4, 1865. Taylor refused to ask for pardon and forfeited the restoration of his citizenship and his plantation. His wife died in 1875, and he moved with his daughters to Winchester, Virginia, where he began work on his *Destruction and Reconstruction*. While visiting a friend in New York, awaiting the publication of the book, he died on April 12, 1879, from a condition believed to have been related to the paralysis which he had in 1861. From the "Introduction" to Richard Taylor's *Destruction and Reconstruction*, edited by Richard B. Harwell, New York, Longmans, Green and Co., 1955.

[3]Bee's division consisted of three more regiments which had not yet made it to Louisiana, but were on the march, the 23rd, 35th, and 36th Texas Cavalry Regiments, Brigadier General Marcus J. Wright, *Texas in the War*, ed. Colonel Harold B. Simpson, Waco, Texian Press, 1955, 3. Hamilton Prioleau Bee was born in Charleston, South Carolina, July 22, 1822, the son of Colonel Barnard E. Bee. Bee had a brother by the name of Barnard E. Bee, who was a general in the Confederate army, being killed at First Manassas after giving "Stonewall"

Jackson his immortal name. Hamilton moved to Texas in 1837 and participated in the Mexican War. He served in the Texas Legislature from 1849 to 1859 and was speaker of the House during the last two years. During 1861 he was in command of State troops, and in March, 1862, he was commissioned a brigadier general and put in command of the Confederate troops at Brownsville. In the winter of 1863-1864 he commanded a force of about 10,000 men from Brazos to Matagorda Bay. In 1864 he commanded a cavalry brigade in the Red River Campaign, participated in the Battle of Mansfield and led an ill-fated cavalry charge on April 9 at Pleasant Hill. After being relieved of command in Louisiana, Bee returned to Texas and was assigned a brigade in Samuel B. Maxey's infantry division. After the fall of the Confederate government he moved to Mexico where he lived until 1876. Bee died on October 2, 1897, and was buried with the Confederate flag (which had been presented to him at the outbreak of the war by the ladies of San Antonio) wrapped around his coffin. Clement A. Evans, *Confederate Military History*, Atlanta, Confederate Publishing Company, 1899. Vol. XI, 225-226. Simpson, editor, *Texas in the War, 1861-1865,* 75-76.

[4]Xavier Blanchard DeBray was born in France in 1819, was graduated from the French Military Academy, and emigrated to Texas in 1852. Settling in San Antonio, Debray published a Spanish newspaper called *El Bejareno*, then moved to Austin, and after the outbreak of the war was Governor Edward Clark's aide-de-camp. In 1862 he was in command of the Military Sub-District of Houston, and commanded the troops on Galveston Island during 1863. Early in 1864 he was ordered to march with his command to join General Richard Taylor in the Red River Campaign. Debray led his regiment at Mansfield on April 8 and the next day at Pleasant Hill. Later he commanded a brigade until his return to Texas at the end of the war. He lived in Houston until 1867 when he moved to Galveston, and finally back to Austin. He died there on January 9, 1895, and is buried in the State Cemetery there near Colonel Buchel and Terrell. Marcus J. Wright, *Texas in the War 1861-1865*, edited by Harold B. Simpson, Hillsboro, Texas, Hill Junior College Press, 1965, 76. Evans, *Confederate Military History*, Vol XI, 226-227.

[5]Richard Taylor, *Destruction and Reconstruction*, ed. Richard B. Harwell, New York, Longmans, Green and Co., 1955, 190.

[6]Robert Selph Henry, *The Story of the Confederacy*, New York, Grosset & Dunlap, 1931, 345; Robert W. Stephens, *August Buchel, Texas Soldier of Fortune*, Privately printed booklet, 1970. Augustus Carl Buchel was born October 8, 1813, at Guntersblum, Germany. He was schooled in the old military tradition of Europe and at eighteen was commissioned a second lieutenant in the First Infantry Regiment of Hesse-Darmstadt, where he served until 1835. Next he served as a lieutenant in the French Foreign Legion, fought in Spain, and was knighted by Queen Maria Christina. Following his service in the Foreign Legion, Buchel served as an instructor in the Turkish army. He returned home after a few years and killed a man in a duel. By law he was re-

quired to leave Germany as the surviving duelist, and after a stay in France, he sailed for Texas in 1845 where he established a home at Indianola. He served Texas in the Mexican War and became an aide-de-camp on the staff of General Zachary Taylor. When the Civil War broke out, Buchel was one of the first to offer his services to his state. He first served in the Texas Militia, but in late 1861 was made lieutenant colonel of the Third Texas Infantry Regiment. In 1863 he became colonel of the First Texas Cavalry Regiment, being ordered to Louisiana in the Red River Campaign.

[7]Major General Tom Green was commander of all the cavalry under General Taylor; Taylor, *Destruction and Reconstruction*, 192. Thomas Green was born in Amelia County, Virginia on June 8, 1814. He was graduated from the University of Tennessee and also from Princeton College, where he had studied law. His father was a Supreme Court judge in Tennessee and was also president of Lebanon Law College. In the fall of 1835, at the age of twenty-one, Green came to Texas and entered the revolutionary army. He fought his first battle at San Jacinto, April 21, 1836, and from then until the disbandment of the army in 1837, he was in most of the skirmishes and engagements of the war. He served as captain of the First Texas Rifle Volunteers during the Mexican War in 1846-1847, and was clerk of the Texas Supreme Court from 1841 to 1861. He entered the Confederate army in 1861, and in August was appointed colonel of the Fifth Texas mounted rifles. Green commanded the land forces in the recapture of Galveston Island on January 1, 1863. In April, 1863, Green was ordered to Louisiana where he fought several battles with the Federals and was promoted to brigadier general on May 20, 1863. He came back to Texas for a short time when it appeared the Federals were going to attempt a campaign up the Rio Grande. Returning to Louisiana, he was promoted to major general and commanded the cavalry under General Richard Taylor. He fought in the Battles of Mansfield and Pleasant Hill and was killed at Blair's Landing April 12, 1864. His body was returned to Texas and he is buried in Austin. Simpson, editor, *Texas in the War 1861-1865,* 78-79. Evans, *Confederate Military History,* Vol. XI, 231-233.

[8]Barr, ed, "Mechling's Journal," April 7.

[9]Barr, ed, "Mechling's Journal," April 7; United States War Department, *War of the Rebellion; A Compilation of the Official Records of the Union and Confederate Armies*, 70 vols. in 128, Washington, Government Printing Office, 1890-1901, XXXIV, Series I, Part 1, 257, 606. (Hereafter this source will be referred to as *Official Records* with the volume and page number. Unless otherwise indicated all references will be to Series I and Part 1.)

[10]*Official Records,* XXXIV, 606.

[11]Thirty-seventh Texas Cavalry: Terrell's Cavalry Regiment, (Also known as the 34th Texas Cavalry), Microfilm Copy 323, The National Archives, National Archives and Records Service, General Services Administration, Washington, D.C., 1961, Rolls 177, 178, 179, from the microfilm in the

Genealogy and History Section of the Dallas Public Library, Dallas, Texas. (Hereafter this source will be referred to as "Microfilm 323, *Terrell's Texas Cavalry*" with the roll number. (Information on other regiments which came from this same source will be shown as "Microfilm 323," with the name of the regiment and roll number following.) It is interesting to note that the officers and men of Terrell's Texas Cavalry always referred to the regiment as the "34th Texas Cavalry" when a number designation was used. Most of the time the men of the regiment simply referred to its name as "Terrell's Texas Cavalry Regiment." There was also a regiment known as the "34th Texas Cavalry (Dismounted)" and had no relationship to Terrell's "34th Texas Cavalry." The 34th Texas Cavalry (Dismounted) had been a cavalry regiment earlier but had been dismounted on November 3, 1862, and trained as infantry. One other regiment at times referred to itself as the "34th Texas Cavalry Regiment." This was the regiment known as "Wells' Texas Cavalry Regiment," which had been formed from remnants of Wells' Texas Cavalry Battalion, Good's Texas Cavalry Battalion, and Gilletts' Texas Cavalry Battalion. Wells' regiment was assigned to the 4th Brigade of John G. Walker's "Greyhound Division" in 1865, Simpson, ed., *Texas in the War* by Marcus Wright.

[12]Lieutenant Colonel John C. Robertson, a 39-year-old lawyer from Tyler, Texas. Born March 10, 1824, he was graduated from Harvard University, where he had studied law. In 1845 he married Miss Sara Goodman and six years later they moved to Tyler, Texas. He was a member of the Texas Secession Convention, chairman of the committee of Public Safety, and was appointed lieutenant colonel of Terrell's Regiment in 1863. After the war he practiced law in Tyler and was elected district judge in 1878 and again in 1880. Robertson died in 1895. Sidney S. Johnson, *Texans Who Wore the Gray*, Tyler, Texas, 1907, 46-47.

[13]Major Hiram S. Morgan from Bastrop, Texas, Microfilm 323, *Terrell's Texas Cavalry*, Roll 178.

[14]Letter from Major General J.B. Magruder to General Cooper, Adj. Gen., C.S., Microfilm 323, *Terrell's Texas Cavalry*, Roll 179.

[15]Special Order No. 145, Hdq. Dist. of Texas, New Mexico & Arizona, *Official Records,* XXVI, Part 2, 26.

[16]*Official Records*, XXVI, Part 2, 99.

[17]*Official Records*, XXVI, Part 2, 106.

[18]S.O. No. 213, Hdq., Texas, New Mexico & Arizona, August 9, 1863, *Official Records*, XXII, Part 2, 963-964.

[19]*Official Records*, XXVI, 975.

[20]S.O. No. 231, Hdq., Texas, New Mexico & Arizona, August 27, 1863, *Official Records*, XXVI, Part 2, 183-184.

[21]S.O. No. 231, Hdq., Texas, New Mexico & Arizona, August 27, 1863, *Official Records*, XXVI, Part 2, 183.

[22]Lieutenant General E. Kirby Smith, commander of the Confederate Trans-Mississippi Department, which included Texas, New Mexico, Arkansas, Missouri, Indian Territory, and Louisiana west of the Mississippi River. He was

born at Saint Augustine, Florida, May 16, 1824, and was graduated at West Point in 1845, entering the Mexican campaign as second lieutenant in the Fifth Infantry. He was promoted to captain in 1855, and engaged in Indian campaigns, reaching the rank of major. He resigned his commission when Florida seceded. In 1861 the Confederate government made him a colonel, and later in the same year, brigadier general. He was severely wounded at the Battle of First Manassas July 21, 1861. He commanded the Confederate forces in the Cumberland Gap region in 1862, after having been made major general. He was made lieutenant general, and finally full general on February 19, 1864, when he was placed in command of the Trans-Mississippi Department. He was the last Confederate general to surrender, May 26, 1865. For a time after the war he was in voluntary exile in Mexico along with Terrell and other Confederate officers. Edmund Kirby Smith died at Sewanee, Tennessee, on March 28, 1893. *The Encyclopedia Americana,* 1956 Edition, New York, Americana Corporation, 1956, 115. Alexander Watkins Terrell, *From Texas to Mexico and the Courts of Maximilian in 1865*, edited by Fannie E. Ratchford, Dallas Book Club, 1933.

[23] *Official Records*, XXVI, Part 2, 223, 535-538.

[24] Lieutenant Colonel Robertson to Colonel A.W. Terrell, September 17, 1863, *Official Records* XXVI, Part 2, 238-239.

[25] *Official Records*, XXVI, Part 2, 278-279.

[26] Lieutenant Colonel Robertson to Captain Edmund P. Turner, assistant adj. gen., December 2, 1863, *Official Records*, XXVI, Part 2, 469.

[27] Charles K. Chamberlain, *Alexander Watkins Terrell, Citizen Statesman*, Ph. D. dissertation, University of Texas, 1956.

[28] *Official Records*, XXVI, Part 2, 281.

[29] *Official Records*, XXVI, Part 2, 469, 470, 471.

[30] *Official Records*, XXVI, Part 2, 454.

[31] Harry McCorry Henderson, *Texas in the Confederacy*, San Antonio, The Naylor Company, 1955, 108.

[32] Microfilm 323, *Terrell's Texas Cavalry*, Roll 177.

[33] *Official Records*, XXVI, Part 2, 563.

[34] *Official Records*, XXXIV, Part 2, 837.

[35] Microfilm 323, *Terrell's Texas Cavalry*, Roll 179.

[36] *Official Records*, XXXIV, Part 2, 988.

[37] *Official Records*, XXXIV, Part 2, 1002.

[38] *Official Records*, XXXIV, Part 2, 1010.

[39] *Official Records*, XXXIV, Part 2, 1011.

[40] *Official Records*, XXXIV, Part 2, 1037.

[41] *Official Records*, XXXIV, Part 2, 1036.

[42] Likens' and Burns' Texas Cavalry battalions were consolidated in late 1863 to form Likens' Texas Cavalry Regiment. The regiment claimed to be the "35th Texas Cavalry Regiment, but this numerical designation was also claimed by Brown's Texas Cavalry Regiment, which was also formed in late 1863. Likens' regiment served in Texas until 1864, when it was transferred to

Louisiana to serve in the Red River Campaign. The regiment participated in the battles of Mansfield and Pleasant Hill in that campaign. Simpson, editor *Texas in the War, 1861-1865,* 121-122. This regiment, along with Yager's (1st Texas Cavalry) was brigaded with Terrell's regiment in September, 1864, in Louisiana, with Colonel Terrell in command. Likens' regiment was dismounted in the reorganization of early 1865, and started for Texas on February 19. See Appendix IV for additional information on Likens' Regiment.

[43]Woods' Texas Cavalry Regiment, commanded by Colonel Peter C. Woods, was formed in late 1863. Its numerical designation was the 36th Texas Cavalry Regiment, and this was apparently the last numerical designation to be assigned by Richmond. The regiment was also called the 32nd Texas Cavalry Regiment. Woods' regiment served in Louisiana. Simpson, editor, *Texas in the War, 1861-1865,* 122. Woods' regiment did not reach Mansfield until April 9, too late to participate in the battles of Mansfield and Pleasant Hill, but was rushed to General Green on April 10, and were a part of the force Green commanded at Blair's Landing where he was killed.

[44]Buchel's Texas Cavalry Regiment was formed from Yager's Texas Cavalry Battalion and Taylor's Texas Cavalry Battalion in 1862. Simpson, editor, *Texas in the War, 1861-1865,* 110-111. After Colonel Buchel was killed at Pleasant Hill, command of the regiment passed to Colonel W.O. Yager, who remained in command for the rest of the war. Yager's regiment was one of the regiments which was assigned to Terrell in September, 1864, as a part of his mounted cavalry brigade. The regiment was not disbanded until May, 1865, at Wild Cat Bluff on the Trinity River with the other two regiments of the brigade, Terrell's 34th Texas Cavalry and Bagby's old regiment, the 7th Texas Cavalry. See Appendix VI for additional information on this regiment.

[45]Debray's Texas Cavalry Regiment was organized early in 1862. Its numerical designation was the 26th Texas Cavalry Regiment, and it was considered to have been one of the best organized and best disciplined regiments in the Confederate army. Its excellent discipline and marching ability often attracted onlookers like a circus passing through town. The regiment served in Texas and in the Red River Campaign. Simpson, editor, *Texas in the War, 1861-1865*, 119. Debray's was one of the regiments in the ill-fated charge of cavalry at Pleasant Hill, where it received the initial shock of the ambushing Federals. In November, 1864, the regiment started for Texas in slow marches, reaching the lower Brazos near Richmond, Texas, at the end of March, 1865. On May 19, Debray's regiment, along with Woods' regiment, made a forced march of fifty miles in fifteen hours to Houston to restore order to that lawless city. The last regimental morning report showed a total of 558 men and 568 horses still present. This was dated May 22, 1865, so apparently the regiment was disbanded on this date, possibly the following day. Thomas H. Edgar, compiler, *DeBray's (26th) Texas Cavalry*, Galveston, Texas, Press of A.A. Finck & Co., 1898, 16-20.

[46]Colonel Nicholas C. Gould commanded Gould's Texas Cavalry Regiment, serving in Texas and the Indian Territory until 1864, when it was sent to

Louisiana. This regiment was the 23rd Texas Cavalry Regiment and like Woods' regiment, did not arrive at Mansfield until the evening of April 9, too late to participate in the battles of Mansfield and Pleasant Hill, but in time to go with General Tom Green in the battle of Blair's Landing. Henderson, *Texas in the Confederacy,* 130.

[47] *Official Records*, XXXIV, Part 2, 1048.

[48] *Official Records*, XXXIV, Part 2, 1096.

[49] *Official Records*, XXXIV, Part 2, 1102.

[50] *Official Records*, XXXIV, Part 2, 1103.

[51] Smith & Mullins, eds., "The Diary of H.C. Medford, Confederate Soldier, 1864," April 3.

[52] *Official Records*, XXXIV, 606, 524.

[53] Nathaniel Prentiss Banks was born at Waltham, Massachusetts, January 30, 1816. Entirely self-taught, he worked himself up from a bobbin boy to become governor of Massachusetts from 1857 to 1859, and in 1861 was appointed major general of Massachusetts volunteers. He fought with credit in the Shenandoah Valley and at Winchester and Cedar Mountain, and at the siege of Port Hudson. He was widely criticized for the failure of the Red River Campaign, a command which he accepted against his better judgment. Years later General Grant placed responsibility for the fiasco on Banks' superiors, but he was never cleared of the charge that he lacked military skill. Banks spent the ten years immediately after the war as congressman from his old district, and was United States marshal for Massachusetts from 1879 to 1888. Banks died at Waltham, Massachusetts, on September 1, 1894. *The Encyclopedia Americana*, Americana Corporation, New York, 1956, 150.

[54] Ludwell H. Johnson, *Red River Campaign,* Baltimore, The John Hopkins Press, 1958, 84.

[55] Robert Selph Henry, *The Story of the Confederacy,* New York, Grosset & Dunlap Publishers, 1931, 341.

[56] Johnson, *Red River Campaign*, 14.

[57] Johnson, *Red River Campaign*, 36.

[58] Johnson, *Red River Campaign*, 93, 94, 100, 112.

[59] Johnson, *Red River Campaign*, 113-116.

[60] Edgar, compiler, *Debray's Texas Cavalry,* 10-11; Taylor, *Destruction and Reconstruction,* 186.

[61] Johnson, *Red River Campaign,* 120.

[62] Johnson, *Red River Campaign,* 124.

[63] Evans, *Confederate Military History*, 130.

[64] Johnson, *Red River Campaign*, 117.

[65] John D. Winters, *The Civil War in Louisiana*, Baton Rouge, Louisiana State University Press, 1963, 338.

[66] Winters, *The Civil War in Louisiana,* 338; Johnson, *Red River Campaign,* 126.

[1]Barr, ed., "Mechling's Journal," April 8; *Official Records*, XXXIV, 606.

[2]*Official Records*, XXXIV, 456.

[3]Taylor, *Destruction and Reconstruction*, 195.

[4]*Official Records*, XXXIV, 607.

[5]Barr, ed., "Mechling's Journal," April 8; *Official Records*, XXXIV, 607.

[6]Taylor, *Destruction and Reconstruction*, 195.

[7]Walker's Texas Division was organized first as "McCulloch's Division," but Major General John G. Walker took command about three months after its organization, and it then became known as "Walker's Division." There were four brigades of infantry at first, but the 4th Brigade was detached from the division shortly after organization, serving east of the Mississippi River. The three remaining brigades included: the 8th, 18th, 22nd, and 13th Texas Infantry regiments in the 1st Brigade; the 11th and 14th Texas Infantry regiments, Gould's Infantry Battalion and the 28th Texas Dismounted Cavalry in the 2nd Brigade; and the 16th, 17th, and 19th Infantry regiments, plus the 16th Texas Dismounted Cavalry in the 3rd Brigade. The division was known as "Walker's Greyhounds" because of the great amount of marching they did. The division saw service in Arkansas and Louisiana, and was present at the battles of Mansfield and Pleasant Hill. The division started for Texas on March 6, 1865, reached Camp Groce on April 15. By April 19 most of the men had left for their homes or were getting ready to do so. Those men who were left in camp were given furloughs on April 20, which was the same as being discharged. Harry McCorry Henderson, *Texas in the Confederacy*, San Antonio, The Naylor Company, 1955, 51-68. John George Walker, son of Missouri State Treasurer John Walker, was born July 22, 1822, in Jefferson City, Missouri. He was commissioned a lieutenant in the U.S. Army in 1846, was promoted to captain in 1847 during the Mexican War, and served in the West and Southwest after the Mexican War. Resigning in July, 1861, he entered the Confederate army as a major of cavalry. Walker was promoted rapidly to lieutenant colonel, colonel, and finally in 1862, to brigadier general. He was at Second Manassas, Harper's Ferry, and Sharpsburg and was made major general on November 8, 1862. Transferring to the Trans-Mississippi Department, he was placed in command of McCulloch's division of infantry and served during the Red River Campaign. Eventually he relieved General Richard Taylor as commander of the District of Western Louisiana. General Walker refused to surrender at the end of the war, and escaped to Mexico and England. He returned to Winchester, Virginia, in the late 1860s and was in the mining and railroad business. Walker was made consul-general at Bogotá, Colombia, later in his life. He died July 20, 1893, in Washington, D.C. John L. Wakelyn,

Biographical Dictionary of the Confederacy, Westport, Connecticut, Greenwood Press, 1977, 424.

[8]Randal's brigade consisted of the 11th and 14th Texas Infantry regiments, Gould's Infantry Battalion, and the 28th Texas Dismounted Cavalry regiment. Harry McCorry Henderson, *Texas in the Confederacy*, 52. Horace Randal was born in Tennessee January 1, 1833, and moved with his family to the vicinity of San Augustine, Texas, in 1839. He attended the United States Military Academy at West Point and was graduated in 1854. After graduation Randal served on the southwestern frontier, resigning from the U.S. Army on February 27, 1861. Refusing an offer of a commission by the Confederates, Randal fought as a private in Virginia during 1861. Upon his return to Texas he was appointed colonel of the 28th Texas Dismounted Cavalry and served in Arkansas and Louisiana during 1862. General E. Kirby Smith recommended him for promotion to brigadier general on November 8, 1863, and his promotion was effective April 8, 1864, the same day he was engaged in the Battle of Mansfield. His brigade was sent to Arkansas after the Battle of Pleasant Hill, and Randal was mortally wounded at Jenkins' Ferry, Arkansas, dying on April 30, 1864. He was buried at Marshall, Texas. Randall County, Texas, bears his name. Simpson, editor, *Texas in the War, 1861-1865*, 89. Evans, *Confederate Military History*, Vol. XI, 251.

[9]Alfred Mouton, or as christened, Jean Jacques Alexandre Mouton, was born at Opelousas, Louisiana, February 18, 1829. His father was governor of Louisiana. Alfred Mouton was graduated at West Point July 1, 1850, but resigned the following September and returned to Louisiana. At the opening of the war he recruited a company from Lafayette Parish, where he was residing, and when the 18th Louisiana was organized he was elected colonel. He served entirely in the West. He was wounded at the Battle of Shiloh, but was made a brigadier general for his conduct in the battle; and when he recovered, was assigned to brigade command in Louisiana. His brigade consisted of the 18th Louisiana and the Crescent Infantry regiments, and Clack's battalion. Later Polignac's Texas brigade was added, and was with him when General Mouton was killed at the Battle of Mansfield leading a charge against the Federals. Evans, *Confederate Military History*, Vol. X, 311-313.

[10]Polignac's Texas Brigade was organized in July, 1862, of three Texas Cavalry regiments, the 22nd, 31st, and 34th. The 20th Texas Cavalry regiment was added to the brigade sometime before October. On about November 1, 1862, the brigade was dismounted and their horses were returned to Texas. The 15th Texas Infantry regiment was added in early 1863, and these five regiments would be retained in the brigade to the end of the war. Polignac assumed command of the brigade in October, 1863, while in Louisiana, and would be in command at the opening of the Battle of Mansfield, as a part of Mouton's division. After General Mouton was killed in this battle, Polignac immediately assumed command of the division. The brigade started for Texas on about March 1, 1865, and on May 24 orders were given to march the troops as near home as possible and discharge them. Alwyn Barr, *Polignac's Texas*

Brigade, Texas Gulf Coast Historical Association, November, 1964.

Camille Armand Jules Marie Polignac was born in France, February 6, 1832. At the beginning of the Civil War he came to America, offered his services to the Confederate government, and was made a brigadier general January 10, 1862, being attached to the Army of Tennessee. Later he was transferred to Louisiana, where he served with great distinction during the Battles of Mansfield and Pleasant Hill. On June 13, 1864, Polignac was commissioned a major general. In 1865, before the end of the war, he returned to France and served there during the Franco-Prussian War of 1870-1871. Subsequently he was engaged in journalism and civil engineering. Evans, *Confederate Military History*, Vol. 10, 314-315. Polignac's son, Prince Victor Mansfield de Polignac, visited Louisiana and the site of the Battle of Mansfield during the Centennial Commemoration Program of the Battle of Mansfield. Plummer, *Confederate Victory at Mansfield*, 55.

[11]Commander of the 28th Louisiana was Henry Gray, who assumed command in mid-1862. Serving entirely within the state of Louisiana, he led the regiment during the difficult times of that state, and was its commander during the Red River Campaign. Before the end of the war he was made brigadier general. He resided in Louisiana after the war and died on December 13, 1892. Evans, *Confedrate Military History*, Vol. 10, 302-304. Gray commanded a brigade of infantry at the Battle of Mansfield, consisting of the 28th Louisiana, 18th Louisiana, and the Crescent Regiment.

[12]The Valverde Battery was organized early in 1862 with the six cannons which had been captured by the Confederates at the Battle of Valverde, New Mexico, in February of that year. The battery served in Arkansas and Louisiana, and did particularly good work during the Red River Campaign. When the war ended, the two 3-inch rifled cannons and the two 6-pounders still assigned to the battery were supposedly thrown into the Red River or buried somewhere near Fairfield, Texas. Two Valverde cannons have been used in celebrations in or around Fairfield in recent years, so it may be that they were actually buried there. Simpson, ed., *Texas in the War, 1861-1865*, 137.

[13]The 2nd Louisiana Cavalry was commanded by an ex-auctioneer by the name of William G. Vincent, and was known as Vincent's Louisiana Cavalry. The regiment fought bravely during the Red River Campaign in that state. Vincent served in the East initially, transferring to the Trans-Mississippi Department sometime in 1862. About September 1, 1862, he was placed in command of the 2nd Cavalry. He was born in Virginia and resided in New Orleans. He was paroled after the war's end at Alexandria, Louisiana, on June 25, 1865. *Records of Louisiana Confederate Soldiers and Louisiana Commands*, compiled by Andrew B. Booth, Vol. III, Book 2, New Orleans, 1920, 526.

[14]Arthur Pendleton Bagby was born in Alabama, and appointed from that state to the United States Military Academy at West Point. He was graduated in 1852, and promoted to brevet second lieutenant of infantry.

Resigning in 1853, he studied law, was admitted to the bar, and practiced at Mobile, Alabama. In 1858 he moved to Gonzales, Texas, and was living there when the war began. During 1861 he was a major in the Seventh Texas, and became colonel of the regiment in 1862. On January 1, 1863, he took part in the Confederate victory at Galveston, Texas, and was promoted to brigadier general in 1863. Just before the end of the war, on May 16, 1865, he was promoted to major general. After the war Bagby practiced law in Victoria, Texas, and was the assistant editor of the Victoria *Advocate*. He died February 21, 1921. Evans, *Confederate Military History*, Vol. XI, 224-225; Simpson, editor, *Texas in the War, 1861-1865,* 75.

[15]Walter Paye Lane was born in Ireland February 18, 1817, emigrating to the United States with his parents in 1821 and settling in Guernsey County, Ohio. He came to Texas in time to fight in the Texas Revolution, and afterwards fought in several Indian battles, including the Surveyor's Fight in Navarro County in 1838. In the Mexican War Lane fought in the battles of Monterrey and Buena Vista. He settled in Marshall, Texas, after the Mexican War and spent a lot of time mining gold and silver in Peru, California, Nevada, and Arizona, making and losing several small fortunes. In 1861 Lane was made a lieutenant colonel and commanded a regiment in the Louisiana Campaign in 1863 and a brigade in the Red River Campaign in 1864, where he was wounded in the Battle of Mansfield. He was soon in the saddle again, and, in October, 1864, was recommended by General E. Kirby Smith for promotion to the rank of brigadier general; his commission was dated March 17, 1865. After the war he returned to Marshall, where he died on January 28, 1892. Simpson, editor, *Texas in the War, 1861-1865*, 85-86; Evans, *Confederate Military History,* Vol. XI, 240-241.

[16]Alonzo H. Plummer, *Confederate Victory at Mansfield*, Mansfield, La., Ideal Printing Co., 1969, 28-29. James P. Major was born in Missouri in 1833. He entered the United States Military Academy in 1852, and was graduated in 1856 as brevet second lieutenant of cavalry, making full lieutenant in December, 1856. In 1857 he served on the frontier, where he was engaged in several Indian battles and skirmishes. In 1859-1861 Major did duty at Indianola, Texas, a commissary depot, and resigned his commission in the United States Army, March 21, 1861, in favor of a Confederate commission the same year. Promotions came fast for him and as colonel he commanded a Texas cavalry brigade in Louisiana with such gallantry that General Taylor urged his promotion to brigadier general. At the Battle of Mansfield he commanded a division of cavalry, fighting as infantry in a magnificent charge that broke the Federal lines. He was in command of a brigade in Wharton's Cavalry Corps in the District of Western Louisiana when the war ended. From 1866 to 1877 he was a farmer in Louisiana and Texas, dying in Austin on May 8, 1877. Evans, *Confederate Military History,* Vol. XI, 245-246.

[17]Taylor, *Destruction and Reconstruction*, 195.

[18]John D. Winters, *The Civil War in Louisiana*, Baton Rouge, Louisiana

State University Press, 1963, 342; Plummer, *Confederate Victory at Mansfield*, 20.

[19]Plummer, *Confederate Victory at Mansfield*, Battle Map on page 28-29.

[20]Taylor, *Destruction and Reconstruction*, 196; Plummer, *Confederate Victory at Mansfield*, 20.

[21]Ludwell H. Johnson, *Red River Campaign*, Baltimore, The John Hopkins Press, 1958, 133.

[22]Winters, *The Civil War in Louisiana,* 342; Johnson, *Red River Campaign*, 134; Taylor, *Destruction and Reconstruction*, 196; Robert S. Weddle, *Plow-Horse Cavalry*, Austin, Madrona Press, 1974, 115; Dudley G. Wooten ed., *A Comprehensive History of Texas,* Dallas, William G. Scarff, 1898, Vol. II, 726.

[23]These were the 83rd, 96th, and 48th Ohio Regiments and the 130th Illinois and 19th Kentucky Regiments. Plummer, *Confederate Victory at Mansfield,* 21.

[24]Winters, *The Civil War in Louisiana,* 343.

[25]Johnson, *Red River Campaign,* 134; Wooten, ed., *A Comprehensive History of Texas,* Vol. II, 726-727; Winters, *The Civil War in Louisiana,* 342; Plummer, *Confederate Victory at Mansfield,* 22; Taylor, *Destruction and Reconstruction*, 354.

[26]J.E. Hewitt, "The Battle of Mansfield, La." *Confederate Veteran,* XXXIII, May, 1925, 172-173. (From the account written by J.E. Hewitt to commemorate the dedication of the monuments to Generals Taylor and dePolignac, April 8, 1925.) H.C. Medford in his diary said that General Mouton had five "mortal wounds," Smith & Mullins, eds., "The Diary of H.C. Medford," 218. This may be true. General W.P. Lane had six holes in his coat after the battle. Two of the bullets brought blood; Walter P. Lane, *Adventures and Recollections of General Walter P. Lane,* Austin, Pemberton Press, 1970, 110.

[27]Plummer, *Confederate Victory at Mansfield,* 22.

[28]Odie Faulk, *Tom Green, Fightin' Texan*, Waco, Texian Press, 1963, 60.

[29]Weddle, *Plowhorse Cavalry*, 116; Alwyn Barr, *Polignac's Texas Brigade,* Texas Gulf Coast Historical Association Publication, Vol. VIII, No. 1, Nov. 1964, 40.

[30]Hewitt, "The Battle of Mansfield, La.," *Confederate Veteran*, 173.

[31]*Annual Report of the Adjutant General of The State of Illinois,* Springfield, Baker & Phillips Printers, 1863, 74.

[32]Winters, *The Civil War in Louisiana,* 343.

[33]Joseph P. Blessington, *The Campaigns of Walker's Texas Division,* New York, Lange, Little & Co., 1875, 188-189; Winters, *The Civil War in Louisiana,* 343.

[34]Johnson, *Red River Campaign*, 135.

[35]Plummer, *Confederate Victory at Mansfield,* 23; Johnson, *Red River Campaign,* 136.

[36]Johnson, *Red River Campaign,* 136; Winters, *The Civil War in Louisiana,* 344.

[37]Smith & Mullins, eds., "The Diary of H.C. Medford," 218; Johnson, *Red River Campaign,* 136-137; Plummer, *Confederate Victory at Mansfield,* 25.

[38]Plummer, *Confederate Victory at Mansfield,* 25.

[39]Johnson, *Red River Campaign,* 137; Winters, *The Civil War in Louisiana,* 344.

[40]Plummer, *Confederate Victory at Mansfield,* 26; *Official Records,* XXXIV, 606.

[41]Johnson, *Red River Campaign,* 138-140; Winters, *The Civil War in Louisiana,* 344.

[42]Plummer, *Confederate Victory at Mansfield,* 26.

[43]Winters, *The Civil War in Louisiana,* 346; *Official Records,* XXXIV, 392, 421-422, 429, 565, 606-607, 616-617.

[44]Blessington, *The Campaigns of Walker's Texas Division,* 191, 201; Smith & Mullins, "The Diary of H.C. Medford," 230; Winters, *The Civil War in Louisiana,* 347; Taylor claimed in his book that 2,500 Federals were captured; Johnson, *Red River Campaign,* 139.

[45]*Official Records*, XXXIV, 564-565.

[46]*Official Records*, XXXIV, 607.

[47]Smith & Mullins, eds., "The Diary of H.C. Medford," 217-218.

[48]Blessington, *The Campaigns of Walker's Texas Division,* Chapter XXIV.

[49]Frank M. Flinn, *Campaigning With Banks in Louisiana, '63 and '64,* Lynn, Mass., Press of Thomas P. Nichols, 1887, 606-609.

Footnotes for Chapter Three

[1]Banks was commander of the Federal forces during the Red River Campaign. See note no. 53, chapter 1, for biographical information.

[2]Johnson, *Red River Campaign*, 146-147. Winters, *The Civil War in Louisiana*, 347. Blessington, *The Campaign of Walker's Texas Division*, 191. Faulk, *Tom Green, Fightin' Texan*, 61.

[3]Blessington, *The Campaigns of Walker's Texas Division*, 191.

[4]Plummer, *Confederate Victory at Mansfield*, 31.

[5]Thomas J. Churchill was born March 10, 1824, near Louisville, Kentucky, and in 1844 was graduated from St. Mary's College. He studied law at Transylvania, and volunteered in the war with Mexico. When the war opened, Churchill was elected colonel of the 1st Arkansas Mounted Rifles. On March 4, 1862, he was commissioned brigadier general, and near the end of 1862 was sent back to Arkansas. Churchill was commissioned major general on March 17, 1863, and was ordered to report to General Bragg in Tennessee; but was

soon transferred to the Trans-Mississippi Department, where he commanded a division of infantry and was in the Battles of Pleasant Hill and Jenkins' Ferry. Evans, *Confederate Military History,* Vol. 10, 394-396.

[6]Mosby M. Parsons commanded a small division of Arkansas infantry with General Sterling Price in Arkansas. This division was detached and sent to General Taylor during the Red River Campaign. General Parsons' division, along with General Churchill's division formed Taylor's right wing at the Battle of Pleasant Hill on April 9, 1864. On the morning of April 14, 1864, Parsons' division was started on the road back to Shreveport to rejoin General Price in pursuit of the Federals in Arkansas. Johnson, *Red River Campaign*, 119, 182.

[7]Taylor, *Destruction and Reconstruction*, 198.

[8]Smith and Mullins, "The Diary of H.C. Medford," 220.

[9]*Official Records*, XXXIV, 566; Plummer, *Confederate Victory at Mansfield*, 34; Johnson, *Red River Campaign,* 153; Faulk, *Tom Green, Fightin' Texan*, 60.

[10]Smith and Mullins, eds., "The Diary of H.C. Medford," 221; Faulk, *Tom Green, Fightin' Texan*, 61; Johnson, *Red River Campaign*, 153.

[11]Wooten, ed., *A Comprehensive History of Texas,* 728.

[12]Taylor, *Destruction and Reconstruction*, 199-200.

[13]Wooten, ed., *A Comprehensive History of Texas 1685-1897*, 728.

[14]Plummer, *Confederate Victory at Mansfield*, 31. General Taylor claimed the Federal forces were about 18,000 in *Destruction and Reconstruction.* Winters says in the *Civil War in Louisiana,* page 355, that the Federals had only about 13,000. Johnson counts 12,193 "effective" in *The Red River Campaign,* page 168. The Confederate force amounted to about 12,000, making the opposing armies very near the same.

[15]*Official Records*, XXXIV, 230; Winters, *The Civil War in Louisiana*, 348-349; Plummer, *Confederate Victory at Mansfield,* 32-33. Johnson, *The Red River Campaign*, 149.

[16]William Polk Hardeman, called "Old Gotch" by the troops, was born in Tennessee on November 4, 1816. He migrated to Texas in 1835 and settled with his father on Caney Creek in Matagorda County. He participated in the Texas Revolution, numerous Indian campaigns, and the Mexican War. His first Confederate service was as a captain in the Fourth Texas Cavalry Regiment in the New Mexico Expedition of 1861-1862. In the latter part of 1862 he was promoted to colonel, and led the regiment in the Red River Campaign. On March 17, 1865, Hardeman was promoted to brigadier general and was in command of a brigade of cavalry in Louisiana. After the war he chose voluntary exile in Mexico, where he was hired to survey lands by the French government. After returning to Texas he engaged in farming until 1874, when he served in various public offices. "Old Gotch" died at Austin on April 8, 1898, and is buried in the State Cemetery near Terrell, Debray, and Buchel. Simpson, editor, *Texas in the War, 1861-1865,* 80-81; Evans, *Confederate Military History*, Vol. XI, 236-237.

[17]This was the 5th Texas Cavalry regiment, formerly commanded by General Tom Green, and now commanded by Colonel Henry C. McNeill. This was the first Texas cavalry unit to arrive in Louisiana. It reported to General Taylor on March 30, 1864, with 250 men, fifty of whom were unarmed. Plummer, *Confederate Victory at Mansfield*, 8.

[18]*Official Records*, XXXIV, 566.

[19]*Official Records*, XXXIV, 566-567.

[20]*Official Records*, XXXIV, 567.

[21]Smith and Mullins, ed., "Diary of H.C. Medford," April 9. (Medford says four men were killed, several wounded, and six horses killed in the Valverde Battery.)

[22]Winters, *The Civil War in Louisiana*, 351.

[23]Wooten, ed., *A Comprehensive History of Texas*, 728; *Official Records*, XXXIV, 567.

[24]Taylor, *Destruction and Reconstruction*, 201-202; Johnson, *Red River Campaign*, 157-158; Chamberlain, *Alexander Watkins Terrell*, Ph.D., Thesis, 74-75.

[25]Johnson, *Red River Campaign*, 160; Taylor, *Destruction and Reconstruction*, 202.

[26]L.A. Pennington, Romeyn B. Hough, Jr., and H.W. Case, *The Psychology of Military Leadership*, New York, Prentice-Hall, Inc., 1943, 223.

[27]Winters, *The Civil War in Louisiana*, 352.

[28]*Alexander Watkins Terrell Papers*, University Archives, The Eugene C. Barker Texas History Center, University of Texas at Austin, Austin, Texas, letter from Lieutenant Colonel John C. Robertson to W.P. Hardeman in 1876 explaining how Colonel Terrell became lost from the command; Chamberlain, *Alexander Watkins Terrell*, Ph.D., Thesis, 74-76.

[29]Winters, *The Civil War in Louisiana*, 352.

[30]Davis Bitton, editor, *The Reminiscences and Civil War Letters of Levi Lamoni Wight*, Salt Lake City, University of Utah Press, 1970, 32, 90; Fredericka Meiners, "Hamilton P. Bee in The Red River Campaign," *Southwestern Historical Quarterly*, LXXVIII, July, 1974, 29. *Official Records*, XXXIV, 608, (Bee's official report of the battle.)

[31]*Official Records,* XXXIV, 608. Thomas H. Edgar, compiler, *History of De Bray's (26th) Regiment of Texas Cavalry*, Galveston, A.A. Finck & Co., 1898, 13; Meiners, "Hamilton P. Bee in The Red River Campaign," 30. There is some confusion as to whether Colonel Buchel died on the battlefield as indicated by Plummer in *Confederate Victory at Mansfield* or a few days later in Mansfield as indicated by Simpson in his notes on page 101 of Wright's *Texas in the War 1861-1865*. Buchel's monument in the State Cemetery in Austin shows his date of death as April 15, 1864, like Simpson. The confusion goes on, however. General Bee says in his official report dated April 10, 1864, at Pleasant Hill, Louisiana, that Buchel's men recaptured him the same day, April 9th, and that he died two days later in his (Bee's) headquarters camp. This, of course, would have been one day *after* Bee's report date of April 10th.

Official Records, XXXIV, 608. Mechling in his journal says that Buchel "fell mortally wounded, since died." It is not likely Mechling was able to write that on the day indicated, April 9. However, it is likely he wrote it the following day, April 10th, which would mean Buchel died on April 10th or possibly April 11th. Barr, ed. "Mechling's Journal," 368. Levi Wight in a letter to his wife dated April 11, 1864, told about Buchel being wounded with three musket balls through him, and that there was no hope for him. He added more to his letter the following day, April 12, saying "Colonel Buchel died last night." However, he mentions in the same April 12 letter that General Green was killed "yesterday evening" so actually Wight's letter should have been dated April 13. This would mean Colonel Buchel died during the night of April 12 or during the early morning hours of April 13. Bitton, ed., *The Reminiscences and Civil War Letters of Levi Lamoni Wight*, 154.

[32]Stephens, *August Buchel*, 5

[33]Barr, ed, "Mechling's Journal," 368.

[34]Hamilton P. Bee, "Battle of Pleasant Hill-An Error Corrected," *Southern Historical Society Papers*, VIII, 1880, 186.

[35] Edgar, compiler, *History of Debray's Texas Cavalry,* 14.

[36]Winters, *The Civil War in Louisiana*, 355.

[37]*Official Records*, XXXIV, 567.

[38]*Alexander Watkins Terrell Papers*, University Archives, The Eugene C. Barker Texas History Center, The University of Texas at Austin, Austin, Texas. Terrell's "Chronology of The War" is a handwritten manuscript depicting his own conception of important events and aspects of the war, much of it based on the official records referred to frequently in this book.

[39]Alwyn Barr, "Texas Losses in the Red River Campaign, 1864," *Texas Civil War Centennial Program for the Centennial Commemoration of the Red River Campaign*, Odessa, Texas, West Texas Office Supply, A Publication of the Texas Civil War Centennial Commission and Texas Historical Survey Committee, 1962, 31.

[40]Smith and Mullins, eds., "Medford's Diary," 223.

[41]One of these bayonets is on display in the museum at the Mansfield battlefield.

[42]*Official Records,* XXXIV, 570-571; Wooten, ed., *A Comprehensive History of Texas*, 730; *Confederate Military History*, X, 1899, 146-147.

[43]William H. Parsons was born in Alabama. He served under General Zachary Taylor during the Mexican War, and went to Texas after it was over, publishing a newspaper at Waco. In mid-1861 he organized and became colonel of the 12th Texas Cavalry at Waxahachie, and served in the Red River Campaign in Louisiana. Parsons was with General Tom Green at the Battle of Blair's Landing, where General Green was killed. Parsons practiced law after the war ended. Simpson, *Texas in the War, 1861-1865,* 115.

[44]*Official Records*, XXXIV, 571.

[45]*Confederate Military History*, X, 146. *Official Records*, XXXIV, 571.

[46]"Account of the Battle of Blair's Landing," *Tom Green Scrapbook*, Texas State Library, Archives Division, Austin, Texas.

[47]Jordan's was a ferry across the Bayou Pierre between Blair's Landing and Pleasant Hill.

[48]"Battle of Blair's Landing," *Tom Green Scrapbook*. *Official Records*, XXXIV, 571.

[49]Faulk, *Tom Green, Fightin' Texan*, 62.

[50]*Riddell Scrapbook, 1864*, 2-1 378, Texas State Library, Archives Division, Austin, Texas. (Sarah Glenn Riddell was Tom Green's sister-in-law.) D.W.C. Baker, *Texas Scrap-Book*, A.S. Barnes & Co., New York, 1875, Texas State Library, Archives Division, Austin, Texas. "Battle of Blair's Landing," *Tom Green Scrapbook.*

[51]"Battle of Blair's Landing," *Tom Green Scrapbook.*

[52]"Battle of Blair's Landing," *Tom Green Scrapbook.*

[53]This is a notation on the bottom of the Muster Roll of Co. B, 32nd Texas Cavalry, in the Texas State Library, Archives Division, Austin, Texas. (The Thirty-second Texas Cavalry was actually Woods' Thirty-sixth Texas Cavalry)

[54]The Galveston *Tri-Weekly News*, April 25, 1864.

[55]"Battle of Blair's Landing," *Tom Green Scrapbook.*

Footnotes for Chapter Four

[1]Barr, ed., "Mechling's Journal," April 15.

[2]*Official Records* XXXIV, 571, 609; *Medford's Diary*, 226.

[3]Barr, ed., "Mechling's Journal," April 20.

[4]*Official Records*, XXXIV, 610; Barr, ed., "Mechling's Journal," April 21, 22.

[5]Johnson, *Red River Campaign*, 224-225. Taylor, *Destruction and Reconstruction*, 235.

[6]Barr, ed., "Mechling's Journal," April 22. General Taylor says in his report that "Terrell's Brigade" was sent to Beasley's Station. Actually there were two regiments sent, Terrell's and Yager's. Apparently Terrell was in command of the two. *Official Records*, XXXIV, 579-580. Meiners, "Hamilton P. Bee in the Red River Campaign," 36.

[7]Barr, ed., "Mechling's Journal," April 24, through April 26.

[8]Barr, ed., "Mechling's Journal," April 27 through April 30.

[9]Barr, ed., "Mechling's Journal," May 2 and May 3.

[10]Edgar, compiler, *DeBray's 26th Regiment of Texas Cavalry*, 16.

[11]*Official Records*, XXXIV, 614. (Bee's report to General W.R. Boggs, chief of staff of the Trans-Mississippi Department dated August 17, 1864.) 585-587.

[12]*Official Records,* XXXIV, 585-587.

[13]Barr, ed., "Mechling's Journal," May 6 and May 7; *Official Records*, XXXIV, 587.

[14]Wooten, ed., *A Comprehensive History of Texas*, 734.

[15]Barr, ed., "Mechling's Journal," May 9.

[16]Johnson, *The Red River Campaign*, 272-274.

[17]Wooten, ed., *A Comprehensive History of Texas*, 735-736.

[18]Edgar, compiler, *DeBray's (26th) Regiment of Texas Cavalry,* 16.

[19]Wooten, ed., *A Comprehensive History of Texas*, 737-738; Johnson, *The Red River Campaign*, 275.

[20]*Official Records*, XXXIV, 588, also Part 3, 368 and 520.

[21]*Official Records*, XLI, 289.

[22]Microfilm 323, *Terrell's Texas Cavalry*, Rolls 177, 179.

[23]*Official Records*, XLI, 289.

[24]Microfilm 323, *Terrell's Texas Cavalry*, Rolls 177, 179.

[25]Not to be confused with the Thirty-fourth (Dismounted) Texas Cavalry who were also operating in the area. There may have been some confusion, but the "Dismounted" regiment was purely infantry, having been dismounted in November, 1862.

[26]*Official Records,* XLI, 854; Terrell also mentions this in his "Civil War Chronology" on file in the *Alexander Watkins Terrell Papers*, University Archives, Barker Texas History Center, University of Texas at Austin, Austin, Texas.

[27]The Dallas *Times Herald*, October 1, 1864. *Official Records*, XLI, 810-812.

[28]*Official Records*, XLI, 810-812, (this was Terrell's official report of the action.)

[29]*Official Records*, XLI, Part 3, 879.

[30]*Official Records*, XXXIV, 580.

[31]*Official Records*, XLI, Part 4, 636.

[32]*Official Records,* XLI, Part 4, 1071. In the same report the First Texas Cavalry (Yager's) regiment was commended for drill, discipline, appearance, police of camp, and care of animals, page 1070.

[33]Winters, *The Civil War in Louisiana*, 387-388.

[34]Winters, *The Civil War in Louisiana*, 414-415.

[35]Davis Bitton, editor, *The Reminiscences and Civil War Letters of Levi Lamoni Wight*, Salt Lake City, University of Utah Press, 1970, 39.

[36]*Official Records*, XLVIII, 1363.

[37]*Official Records*, XLVIII, 1396.

[38]*Official Records*, XLVIII, 1351; Barr, *Polignac's Texas Brigade*, 53-54.

[39]*Official Records*, XLVIII, 1352, 1390.

[40]Bitton, ed., *Reminiscences and Civil War Letters of Levi Lamoni Wight*, 176, *Official Records*, XLVIII, 1390.

[41]Bitton, ed., *The Reminiscences and Civil War Letters of Levi Lamoni Wight*, 178. Wright confirms that Terrell's (37th), 1st, and 7th Cavalry

regiments were in a brigade, see page 3; Wright, *Texas in the War, 1861-1865*, Simpson, ed., *Official Records*, XLVIII, 1390.

[42]Bitton, ed., *The Reminiscences and Civil War Letters of Levi Lamoni Wight*, 93.

[43]Winters, *The Civil War in Louisiana*, 418-419.

Footnotes for Chapter Five

[1]Evidence exists which indicates that Terrell's entire brigade of three regiments was present and disbanded at Wild Cat Bluff. B.F. Broyles was a lieutenant in the 7th Texas Cavalry, which had been part of Terrell's brigade in Louisiana, and he indicates in *The Lone Star State* that his regiment also disbanded at "Wildcat on the Trinity." Broyles' military record is found on Microfilm 323, *7th Texas Cavalry*, Roll 45.

[2]Winters, *The Civil War in Louisiana*, 422.

[3]Stephen B. Oates, *Confederate Cavalry West of the River*, Austin, University of Texas Press, 1961, 156-158.

[4]Alexander Watkins Terrell, *From Texas to Mexico and the Courts of Maximilian in 1865*, edited by Fannie E. Ratchford, Dallas Book Club, 1933, 1-5.

[5]Wild Cat Bluff had a post office established on December 12, 1859, and it was operated there until November 5, 1866, when it was discontinued. It was reestablished July 23, 1869, but discontinued again on January 12, 1870. Mr. E.R. Whatley was the first postmaster and he also operated the ferry across the Trinity River. The ferry continued in operation until the 1930s. Michael J. Vaughn, *The History of Cayuga and Cross Roads, Texas,* Waco, Texian Press, 1967, 9, 88, 99.

[6]Terrell, *From Texas to Mexico in 1865*, 1-5. Terrell does not give the date in the published book, but on the original manuscript of the book on file in the University Archives, "May 15" was shown at first and later crossed through, omitting it from the published book.

[7]Winters, *The Civil War in Louisiana,* 422-423.

[8]*Official Records*, XLVIII, Part 2, 1307.

[9]Oates, *Confederate Cavalry West of the River*, 160.

[10]Winters, *The Civil War in Louisiana*, 425.

[11]Terrell, *From Texas to Mexico in 1865*, 15.

[12]Terrell, *From Texas to Mexico in 1865*, 5-8.

[13]Terrell, *From Texas to Mexico in 1865*, 9-11.

[14]Terrell, *From Texas to Mexico in 1865*, 15; General King said later in a letter to Terrell dated November 3, 1906, that he forgave Terrell on the spot, *Alexander Watkins Terrell Papers*, University Archives, Barker Texas History Center, Austin, Texas.

[15]Terrell, *From Texas to Mexico in 1865,* 17.

[16]Terrell, *From Texas to Mexico in 1865,* 18-19.

[17]Terrell, *From Texas to Mexico in 1865*, 25-33.
[18]Terrell, *From Texas to Mexico in 1865*, 45-46.
[19]Terrell, *From Texas to Mexico in 1865*, 51-63.
[20]Terrell, *From Texas to Mexico in 1865*, 72, 80.
[21]Terrell, *From Texas to Mexico in 1865*, 81, 87-88.
[22]Terrell, *From Texas to Mexico in 1865*, 32.

APPENDIX

Appendix I

Alexander Watkins Terrell

Early Life

In 1832 a young Virginia physician by the name of Christopher Joseph Terrell arrived in Missouri with his wife and sons, caught up in the burning desire of the times to go West. Dr. Terrell purchased two small farms, one about two miles from Boonville, Missouri, and the other about twelve miles. The family lived in town in Boonville, where Dr. Terrell practiced medicine. He was not to live long in his new adopted state. Cholera claimed his life on August 18, 1833, at age 36, leaving a wife and his three sons, Alexander Watkins, John J., and James C. His last wish was that his wife, Susan, see that the sons received a good education. Susan Terrell did just that. John became a doctor and moved back to Virginia and Alexander and James became lawyers. James would later move to Texas, too, and make his home in Fort Worth.[1]

Alexander was born on November 3, 1827,[2] in Patrick County, Virginia. After his father died, when Alexander was five years old, his mother sold the Boonville home and moved to the farm near Boonville. The sons did the farm work after school and on Saturdays—often plowing by moonlight. They hunted and learned how to live in the woods. Alexander became a crack shot with a rifle and in deer drives he was often given the choice stand from which he could shoot the deer.

Terrell considered himself a poor student, confessing a preference for hunting and fishing to study. However, although he considered himself a poor student, he was well advanced in mathematics, Latin, and Greek before he went to the university. He had read many translations of the classics, including Homer's *Iliad,* Plutarch's *Lives,* Cicero's works, and Gibbon's *Decline and Fall of the Roman Empire.*

ADMITTED TO THE BAR

After finishing the primary schools and a course at the University of Missouri, Terrell took up the study of law in 1847 at Boonville with Judge Peyton R. Hayden. He was admitted to the bar in 1849 and began practice at St. Joseph, Missouri, until September 15, 1852. On that date Terrell and his young wife, Ann, left Missouri and headed for Texas with their two girl babies. Ann Terrell had consumption and they hoped a warmer climate would help her.

TO TEXAS

Loading all their possessions on a single wagon, pulled by four mules, and with a saddle horse trailing behind, they headed southward toward Austin, reaching that city on November 2, 1852.

Lawyer Terrell practiced law in the Austin area until 1857 when he was elected judge of the Second Judicial District, and served in that capacity until his commission in the Confederate cavalry.[3]

In the third year of his term as district judge, on July 16, 1860, Ann Terrell died at the age of thirty, leaving Terrell with five children. Terrell had hoped for the eight years since moving to Texas that Ann would be restored to health. For four months before her death he hardly left his house, staying with her constantly. Often he had Ann's bed moved out into the yard so she could get sunshine. His court was simply not opened during this time, and the lawyers did not expect it to be, urging Terrell to stay with his wife.

TERRELL'S CONFEDERATE SERVICE

When Terrell's term as district judge expired in 1863, he joined the Confederate army. There is much evidence that he had been active in the army as a volunteer aide-de-camp as early as 1861. Apparently he was able to work the court calendar to the benefit of the Confederacy during this time, serving both the army and the court.

Terrell thought that the Civil War was not necessary. He felt that a compromise could have satisfied both the North and South, and he was a close friend of Sam Houston, who was strongly against secession. Most of the people in Austin were against secession. The vote there was 704 to 450 against secession,[4] and due to his association with Sam Houston, Terrell probably voted against it with the majority. He took no part in the movement to secede.[5]

On March 24, 1862, Governor Francis R. Lubbock recommended Terrell for a commission, saying that Terrell had served six years as district judge and had served " . . . with the Arkansas Army without pay or rank, which was declined by him several times for nearly the year passed . . . " Lubbock said Terrell had one of the "finest minds in the state, was young, athletic, and energetic."[6] Terrell was named on the staff of Brigadier General H. E. McCulloch as volunteer aide-de-camp on June 12, 1862.[7] He was still on the staff as of December 24, 1862.[8] On November 6, 1862, Terrell, now a captain, purchased cloth at Little Rock, Arkansas, for General McCulloch's command. A letter to Terrell on February 17, 1886, from a G. W. Jones praised Terrell for his duty at the time of his Arkansas service, reminding Terrell that he, Jones, had met Terrell one night at Camp Nelson, Arkansas, near General McCulloch's tent.[9]

Terrell considered joining the Confederate army in 1861. On November 4 of that year he wrote to Brigadier General P. O. Hebert at Galveston accepting a commission which had been offered to him as colonel of the Eighth Regiment of infantry. He apparently had some second thoughts about this, because on November 28 he wrote Hebert that because of the difficulty in raising twelve months' troops, he must decline the commission.[10]

President Davis was sent a letter on March 12, 1863, by G. C. Hebert and others recommending Terrell for appointment as a Brigadier General of Texas troops. The letter highly praised Terrell saying that he had "high character as a man and integrity and ability as a judge" and "a generally conceded aptitude for military position."[11] No record is shown as to the action taken on this letter by President Davis or the secretary of war. However, on April 28, 1863, another letter was written by Brigadier General William R. Scurry, commanding the East Sub-District of Texas, to Lieutenant General Smith, commanding the department at Shreveport, recommending Terrell for lieutenant colonel of cavalry. The letter stated in part that Terrell was a "gentleman of most excellent talents, education, of great energy, unquestionable moral habits and of that discipline of mind which would make a most efficient active . . . officer."[12]

TERRELL BECOMES COMMANDER OF CAVALRY REGIMENT

Terrell was finally to find his niche in the Confederate service. On March 31, 1863, Major General Magruder commissioned him a lieutenant colonel "to command one of the battalions to be raised for service in Arizona, New Mexico and now to be employed in repelling the attacks of the Indians and parties from Kansas who are committing outrages on the people of North Western Texas." The battalion was ordered to Bonham, Texas, on May 30, 1863, but these orders were changed and the battalion was increased to a regiment on June 8, 1863, at Navasota, Texas. Formal organization occurred on June 20, 1863, and Terrell was made colonel of the regiment.[13] The regiment moved south about twenty miles to Hempstead on June 30, where there was a training facility by the name of Camp Groce.[14]

Terrell served gallantly and bravely during the Red River Expedition. An article in the Galveston *News* in July, 1893, said that the author (W. P. D.) was an eyewitness to the bravery of Terrell on several occasions during this expedition.[15] Sinclair Moreland in *The Home and State* on September 28, 1912, says Terrell was a "daring and efficient officer."[16] The conditions of battle during the Civil War were such that officers were required to actually lead their men into battle. It was necessary for the officers to set an example of coolness and courage, and present the appearance of bravery in a calm, controlled manner. In all the charges made against Federal lines during the Red River Campaign, Terrell would have been compelled by the custom of battle of the times to lead, or be at the front of his troops during the charge. There is no doubt that he did so. That he was not killed was truly against the odds for an officer of his rank. Both Federal and Confederate soldiers made it a special point to attempt to shoot the officers of the enemy. W. P. Lane had six bullet holes in his coat at Mansfield, General Mouton had five. Colonel Buchel at Pleasant Hill had three Minié balls through his body.[17]

TERRELL IN MEXICO

Terrell and his group of Confederate officers reached Monterrey, Mexico, where they were joined by still others. On July 4, 1865, ex-General John B. Magruder presided at a banquet given for about fifty such officers of the Confederacy. Generals W.H. King, Bee, Magruder, Hindman, Hardeman, Cadmus Wilcox, Walker, and Terrell were present. Also present were Colonels E. P. Turner, Flournoy, and Duff. Toasts were drunk to the "Lost Cause," to Lee, to Jackson, and to the sovereignty of the states.[18]

Later, Terrell was appointed as *chef d' bataillon* in the French army of occupation, a position equivalent to the rank of colonel in the Confederate or United States armies. Napoleon III of France had sent troops into Mexico and occupied it under the government of Maximilian during the American Civil War. Now that the war between

North and South was over, the United States turned its attention to the situation in Mexico and demanded that Napoleon withdraw. In January, 1867, the French troops did withdraw, leaving Maximilian with only a small force. This force was quickly defeated by the Mexicans and Maximilian was shot by a firing squad. Terrell had anticipated that this was going to happen and had resigned his French commission, returning to Texas in late 1865, having served in the French army only about four months from August through November.

TERRELL RETURNS FROM MEXICO

Terrell settled briefly in Houston, practicing law. He soon became disgusted with the domination of the courts by military governors of the Reconstruction period and became a cotton planter on his farm on the Brazos River in Robertson County. His second wife, formerly Miss Sallie D. Mitchell, died in 1871 and he moved back to Austin that same year.[19] Here he began to practice law again and in 1874 was elected reporter for the Supreme Court of Texas. In 1875 he was elected to the State Senate and distinguished himself in legislation to correct some of the wrongs of Reconstruction. Subsequently he served terms in both branches of the Texas Legislature and authored such bills as the Terrell Election Law, improvement in the jury system, the State Railroad Commission, the bill providing for the construction, without taxation, of the capitol building, and many other bills. He was the joint author of the bill organizing the state university, and later served on the board of regents of that institution. His efforts in behalf of the state university earned him the title of "Father of the University of Texas."[20]

In 1883, Terrell was married to Mrs. Ann H. Jones, formerly Miss Ann H. Holliday, and continued to make his home in Austin.

Terrell had a deep voice, pleasant to listen to, and was a very entertaining and impressive public speaker. His quick mind, deep voice, and athletic physique always commanded respect. He was looked upon as a champion of the people and had many interests as well as his work, particularly in the literary field.[21]

One of his literary efforts was a poem which he wrote shortly after the death of John Wilkes Booth and which probably accurately illustrates the inflamed spirit of many people at this time. The poem was written at a time when there were rumors across the land that

Booth's body was taken out to sea and buried there. It was published anonymously, but it is known now that Terrell was the author. Since then it has been published several times as evidenced by the different copies of it in the University of Texas archives. In April, 1913, it appeared in the *Confederate Veteran* and the following is taken from that publication. It is presented here, not as an indication of Terrell's feeling but to endeavor to show his literary versatility.

"Give him a sepulcher
Broad as the sweep
Of the tidal wave's measureless motion,
In the arms of the deep
Lay our hero to sleep,
'Mid the pearls of the fetterless ocean,

It was liberty slain
That so maddened his brain
To avenge the dead idol he cherished.
So 'tis meet that the main,
Ne'er curbed by a chain,
Should entomb the last freeman now perished.

For the dust of the brave
Could not rest in the grave
Of a land where blind force hath dominion.
Then give him a grave
Underneath the blue wave,
Which the tyrants of earth cannot pinion,

He who dared break the rod
Of the blackamoor's god,
All the host of the despot defying,
Could not rest 'neath the sod
That his minions had trod,
Who was shamed by his glory in dying.

Then hide him away
From the sad eye of day,
'Mid the coral of sea-green abysses,
Where the mermaids so gay,
As they sport 'neath the spray,
May purple his pale lip with kisses,

As the ocean streams roll
From the gulf to the pole,
Let them mourn him with musical dirges;
And the tempest shall toll
For the peace of his soul,
More sublime than the sound of its surges,

He has written his name
In bright letters of fame
In the pathway of liberty's portal;
And the serfs who now blame
Will crimson with shame
When they learn they have cursed an immortal,

He hath died for the weal
Of a world 'neath the heel
Of too many a merciless Nero;
But while yet there is steel,
Every tyrant shall feel
That God's vengeance but waits for its hero.

Then give him a sepulcher
Broad as the sweep
Of the tidal wave's measureless motion;
In the arms of the deep
Lay our Brutus to sleep,
Since his life was as free as the ocean.''[22]

TERRELL GOES TO TURKEY AS MINISTER

During President Cleveland's second term of office, Turkey was a victim of a reign of terror as the Armenians and Turks were engaged in bloody conflicts and deadly encounters. Many missionaries in Turkey were being massacred along with those engaged in the conflict, having chosen one side or the other. President Cleveland realized that he needed a man in Turkey who was intelligent and clever, and for this job he chose Alexander Terrell. He was sent to Turkey as minister to the royal court of Abdul Hamid and served from 1893 to 1897. Terrell became friends with the sultan and as a result no American missionaries were injured, although many from other countries were tortured and killed.

TERRELL'S DEATH AND BURIAL

In 1912 Terrell visited his relatives in Virginia and on the return trip to Austin stopped off at the Crazy Water Building, a hotel in Mineral Wells, Texas. One of his last letters was written here on September 8, 1912, to his granddaughter, Constance James, in Austin.[23] On September 9 he went automobile riding with some friends, returning to the hotel in the early afternoon, feeling well and unusually cheerful. A few hours later he was found dead in his room, apparently of a heart attack.

Terrell was buried in the State Cemetery in Austin with other heroes of Texas. The funeral was on September 12, 1912, at the First Presbyterian Church. A tune "Serenity" was played and songs, "O God Our Help in Ages Past" and "Lead Me Gently Home" were sung. At the grave "Softly Now the Light of Day" was sung.[24]

Today, Terrell's grave rests on a beautiful tree-covered hill near other heroes of Texas and near his old comrades during the Red River Campaign, W.P. Hardeman, Xavier B. Debray, and August Carl Buchel. Below the hill are buried many of the men who served with them during the trying time in Louisiana.

TERRELL'S PLACE IN HISTORY

Alexander Watkins Terrell was regarded as one of Texas' greatest patriots and statesman. He was unselfish and charitable, always ready to sacrifice his own interests to that of his country. A life-size oil painting of him hangs in the capitol in Austin, along with the other great men in Texas history. He was considered to have authored "more good laws for Texas than any other man living or dead."

FOOTNOTES

[1]Chamberlain, *Alexander Watkins Terrell, Citizen, Statesman,* Ph.D. dissertation, 3-5.

[2]Frank W. Johnson, *A History of Texas and Texans,* III, The American Historical Society, Chicago, 1914, 1063. Terrell's tombstone in State Cemetery in Austin shows his date of birth as November 23. Chamberlain in his Ph.D. dissertation concludes that it was probably November 3.

[3]*Confederate Veteran*, "Judge Alexander Watkins Terrell," XX, December 1912, 575; Chamberlain, *Alexander Watkins Terrell, Citizen,*

Statesman, Ph.D. dissertation, 6-10; Mary Ella Wallis *The Life of Alexander Watkins Terrell,* 1827 to 1912, M. A. thesis, University of Texas, Austin, Aug., 1937 (Microfilm in Dallas Public Library, Dallas, Texas).

[4]Chamberlain, *Alexander Watkins Terrell, Citizen, Statesman,* Ph.D. dissertation, 57-60.

[5]Johnson, *A History of Texas and Texans,* 1063-1064.

[6]Microfilm 323, *Terrell's Texas Cavalry,* Roll 179.

[7]*Official Records,* IX, 718.

[8]*Alexander Watkins Terrell Papers,* The Eugene C. Barker Texas History Center, University Archives, The University of Texas at Austin, Austin, Texas. (From Terrell's own "Chronology of the War," a handwritten chronology of the events of the war.)

[9]*Alexander Watkins Terrell Papers*, University Archives, Barker Texas History Center, 2H7.

[10]Microfilm 323, *Terrell's Texas Cavalry,* Roll 179.

[11]Microfilm 323, *Terrell's Texas Cavalry,* Roll 179. This letter is rather faded and it is difficult to read the signatures. The "G. C. Hebert" may have been "P. O. Hebert."

[12]Microfilm 323, *Terrell's Texas Cavalry Regiment*, Roll 179.

[13]Microfilm 323, *Terrell's Texas Cavalry,* Roll 179; *Official Records*, XXVI, Part 2, 73.

[14]Apparently this camp was on or near the plantation owned by Jared E. Groce, a Virginian, who, after operating large plantations in South Carolina, Georgia, and Alabama, brought his slaves and moved to the Brazos River in 1822. Lewis W. Newton and Herbert P. Gambrell, *A Social and Political History of Texas,* Dallas, The Southwest Press, 1932, 109-110.

[15]*Alexander Watkins Terrell Papers,* University Archives, Barker Texas History Center, 2H18, (Galveston *News*, July 16, 1893).

[16]*Alexander Watkins Terrell Papers,* University Archives, Barker Texas History Center, 2H20.

[17]Bitton, ed., *The Reminiscences and Civil War Letters of Levi Lamoni Wight*, 153.

[18]Terrell, *From Texas to Mexico and the Court of Maximilian in 1865,* 21.

[19]L. E. Daniel, *Personnel of the Texas State Government,* San Antonio, Maverick Printing House, 1892, 315.

[20]Sinclair Moreland, "Life Sketch of A. W. Terrell," *The Home and State,* Dallas, Texas, Vol. 14, No. 12, September 28, 1912; *Confederate Veteran,* XX, December, 1912, 575-576.

[21]Daniel, *Personnel of the Texas State Government,* 316-319.

[22]*Alexander Watkins Terrell Papers,* The Eugene C. Barker Texas History Center, University Archives, The University of Texas at Austin, Austin, Texas, 2H18; *Confederate Veteran,* April 1913.

[23]*Alexander Watkins Terrell Papers,* University Archives, Barker Texas History Center.

[24]The Austin *Statesman,* September 10, 1912.

Appendix II

Sketches of Terrell's Cavalry Regiment

Companies

COMPANY A

Company A consisted of men mainly from Prairieville, Texas, and a few from Kaufman and Canton. These were men primarily of Norwegian descent. Six foot, four inch Israel Spikes was the original captain, being elected on December 15, 1862, when the company was formed. Spikes resigned about a year later and William G. Hill became captain.[1] Israel Spikes was a committeeman in 1875 to help write the Texas constitution, and the auburn haired "Swede" was called the "Big Swede" by his friends because he was an arbitrator and peacemaker for the Norwegians and had many friends among them.[2]

The discipline, instruction, arms, and clothing of Company A were rated as "good" on the February 29, 1864, company report. Military appearance was shown as only "fair," and accouterments as "good," but without bayonet scabbards.[3]

At the disbandment in May, 1865, at Wild Cat Crossing on the Trinity River these men were only a short distance from home. Many of the men moved to the Clifton, Texas, area after the war and started new homes and lives.

COMPANY B

This company was organized on January 10, 1863, at Greenville, Texas, and most of the men were from that area. George W. Cooper was the original captain. The thirty-eight-year-old druggist led the company through its early days in Texas and the bloody battles of Mansfield, Pleasant Hill, Lecompte, and Yellow Bayou, Louisiana. He resigned his commission January 10, 1865, having served exactly two years with the regiment. Lieutenant Prior Hart took command of the company upon Cooper's resignation and stayed with it till the end at Wild Cat Bluff.[4]

Company B had "good" discipline, military appearance, arms, and clothing, "good" accouterments, but with no bayonet scabbards, and only "limited" instruction, the latter being primarily because of the company's detached service.[5]

DETACHED SERVICE OF COMPANY B

On August 9, 1863, Major General J. B. Magruder ordered Company B to Waco and vicinity on temporary duty to clean out nests of deserters in the area who had become a terror to the citizens there. The company made camp near Belton on August 25 and immediately began the work of tracking down the deserters. By September 2 only one deserter had been caught, but several persons had been arrested who were known to be aiding and feeding them.

Leader of the deserters was a man by the name of Gelisha Bivins. Twenty-eight well armed men were with him, living in the woods and taking whatever they wanted from the people in the area. These men fled the Belton area when Captain Cooper arrived with Company B, joining a still larger group of forty deserters in Hamilton County. This group was led by a man named Gholstone.

Many of the deserters arrested were from Gurley's regiment and some were given arms and sent back into active Confederate service. While in the vicinity, Captain Cooper raised two companies of men and sent them to W. Baylor's regiment. Company B also received twenty new recruits from Hunt County for duty. Indians were also active in the area and the presence of Company B did much to ease the minds of the local citizens.

On January 3, 1864, the company was in camp fifteen miles west of Waco when orders were received to cease activity in that area and report at once to the regiment. Moving out on about January 10, the company marched 200 miles with horses and wagons, joining the regiment at Camp Dixie, Texas. The regiment was now complete again.[6]

COMPANY C

Organized January 31, 1863, most of the men came from Athens and Waxahachie. Captain was William Preston Payne, a farmer from Waxahachie. Discipline and instruction of this company were reported as "very good." Arms and clothing "good," and accouterments "good," but with no bayonet scabbards. Military appearance was only "fair." Wild Cat Bluff was only a few miles from the homes of most of these men.[7]

COMPANY D

This company was organized by George Washington Diamond, the men enlisting on various dates in 1863, mostly in March and April of that year. Diamond initially had enlisted as a private in May, 1861, transferred to the 11th Texas Cavalry regiment after the "great hangings" at Gainesville in 1862. Diamond wrote an account of the hangings in Gainesville, which was not published until many years later. After the war was over, Diamond returned to Henderson and was subsequently elected state representative. However, the Reconstruction of Texas by military forces did not allow this legislature to convene until 1870 and by this time Diamond had moved with his family to Whitesboro, where he lived until his death on June 24. 1911.

Diamond practiced law in the county seat of Sherman during the Reconstruction, held several public offices, and was on the staff of the Whitesboro *News.*[8]

Company D shared the plight of most of the other companies in Terrell's regiment—they had "good" accouterments, but no bayonet scabbards. Discipline, instruction, arms, and clothing were all "good," but military appearance was only "fair."[9]

Lieutenant Irving P. Mason of Company D and six privates were captured in the battle of Pleasant Hill. The records show that Lieutenant Mason was exchanged on April 20, 1864, at Blair's Landing, where General Green had been killed.[10] There were not many casualties in the Battle of Mansura on May 16, 1864, but there was at least one. Sergeant J. H. Cuningham of Company D was killed during this affair.[11]

COMPANY E

Palestine, Texas, was the home of most of the men of Company E, not far from Wild Cat Bluff. It was organized on April 11, 1863, by Captain Reuben A. Reeves of Palestine. Reeves resigned in September, 1864, to take his newly elected position as associate justice of the Supreme Court of Texas.[12] Discipline was "very good," instruction, arms, accouterment (except for scabbards), and clothing was "good." Military appearance was "fair."[13] Lieutenant Nathan C. Gunnels had an arm broken in the Mansfield or Pleasant Hill battles and Private R.H. Gilmore was wounded slightly in one of the same battles, but Private Elisha Maine was killed.[14] Private R. C. McKenzie was wounded in the left side on May 13, 1864.[15] Private John T. Rice was captured near Morganza, Louisiana on August 8, 1864.

COMPANY F

This company was organized in April, 1863, with most of the men coming from Carthage. Captain John K. Williams apparently stayed with the company to the end. The February 29, 1864, report did not mention a shortage of bayonet scabbards, with all aspects being reported as "good," except for military appearance, which was "fair."[16] This company had four privates missing in the Mansfield-Pleasant Hill battles, and no doubt many were later found dead, or were captured.

COMPANY G

Nacogdoches was the home of most of the men in this company when it was organized in April, 1863, although there is one notable exception. Lieutenant Thomas J. Waller was from Michigan, where, he mentions in a letter, he was compelled to leave because of his pro-Southern sentiments. A twenty-six-year-old editor by the name of Samuel H. B. Cundiff was captain throughout the life of the company. Lieutenant John F. F. Doherty walked into the Federal lines and gave himself up some time before April 21, 1864. He signed the oath of allegiance to the United States on April 25, 1864. The company had several casualties in the Mansfield-Pleasant Hill battles. Sergeant M. J. Davis was killed in action and Privates Thomas J. Jennings and William B. Moore were both wounded.[17] Sergeant D. M. McKnight and Private Eli A. Day were wounded on May 13 and Private Isaac N. Pike was wounded on May 15.[18] Private Oscar L. Davidson was captured September 11 and sent to Elmira, N. Y., where he died on December 12, 1864.[19]

This was the only company in the regiment which had a "superior" rating on instruction. Its rating on all other factors, i.e., discipline, military appearance, arms, accouterments, and clothing was "good."[20]

COMPANY H

Company H rated "good" on all military factors. Captain James F. Warren was company commander. The majority of the men came from Tyler or Rusk. This was the company that was with Colonel Terrell when they were cut off by the Federals during the battle of Pleasant Hill. Several were wounded in the Mansfield-Pleasant Hill battles: Lieutenant James T. Garrett, Corporal Benjamin F. Stamps, and Private Andrew J. Praytor. Private Alex H. Goodman was killed in action.[21] Lieutenant Gabriel D. Gilley and Privates Ballard A. Day, James L. Felton, and William Zuber were all captured near Morganza, Louisiana on August 25, 1864. The latter three were sent to the infamous prison at Elmira, N.Y. Felton and Zuber[22] both died there under conditions comparable to those in the equally infamous Andersonville prison in Georgia. The commissary-general of prisoners, Colonel W. Hoffman, described Elmira as " . . . a festering mass of corruption, impregnating the entire atmosphere of the camp with its pestilential odors, night and day," and "filling the air with its messengers of disease and death, the vaults give out their sickly odors and the hospitals are crowded with victims for the grave."[23] This company once reported itself "Prepared to do good service and ready to do it."[24]

COMPANY I

Originally Company I was organized on April 27, 1863, with Captain C. G. Murray as its commander. However, after the mutiny in September, 1863, the company — what was left of it — was merged with Taylor's company and became Company K. A new Company I was formed under the command of Captain Paschal R. Turner. This company was reported on February 29, 1864, as "good" in discipline, instructions, arms, and clothing, "good" on accouterments, except for lack of bayonet scabbards, and "fair" on military appearance. Most of the men were from Bastrop and Columbus.[25]

Sergeant James M. Reding, Private H.W. Jones, and Private Ellison Waddle were wounded during the Mansfield-Pleasant Hill battles.[26] Although no record was found, Private John M. Denson was probably wounded also, as he is listed as being in the hospital at Shreveport on June 30, 1864, remaining until July 6, 1864, when he was transferred to a hospital in Keatchie.

Corporal George Reager died in the Elmira prison camp in New York after having been captured near Morganza, Louisiana, on August 25, 1864. Private K.H. Barbee was captured along with Corporal Reager and was one of a group of sick Confederate prisoners paroled before the end of the war. No doubt he was one of those whose condition was extremely bad.

On June 21, 1864, Captain Murray, who was apparently still in confinement for his part in the August mutiny of the preceding year, wrote a letter to General Kirby Smith stating in part that when the order to dismount was given, it caused general discontent and the men refused to obey, saddled and rode off, and he went with them for the sole purpose of getting them into the service again. The records do not show the final disposition of Captain Murray's case. Perhaps he was released to fight in the Red River Campaign as Texas needed every able-bodied man.[26]

COMPANY K

The original Company K was organized on April 1, 1863, but was reorganized after the mutiny and its commander became Captain Russell J. Starr. Smith, Cherokee, Upshur, Wood, and Van Zandt counties were home to most of the men. Captain Starr apparently was the commander throughout the Red River Campaign. His wound in the hand on May 14, 1864, was not serious enough to send him to the hospital.[28] Lieutenant Jessie G. Chancellor, who had been one of the leaders of the mutiny in August, 1863, was still confined in the guardhouse in Galveston on June 28, 1864, but he signed his parole at Marshall, Texas, on August 18, 1865, as second lieutenant of Company D. It is possible, therefore, that he was returned to duty with the regiment in Louisiana after Lieutenant Mason of Company D resigned on May 27, 1864.

Sergeant W.O. Weisner was killed in action in the Mansfield-Pleasant Hill battles, and at the same time Corporal Samuel L. Scott was wounded. Seven privates were missing, many apparently killed in action.[29] Lieutenant Robert W. Spradling was wounded in the fight on May 13, 1864,[30] and was sent to the hospital in Keatchie. Private George W. Churchwell was wounded somewhere in Louisiana, and was in the hospital at Keatchie, Louisiana, on April 17, 1864.[31]

Company K seems to be one of the better companies in the regiment, along with Company G and Company H. Company K rated "good" in all areas on the February 29, 1864, report.[32]

REGIMENTAL STRENGTH

The aggregate strength of the regiment on February 29, 1864, was 734, forty-six commissioned officers and 688 enlisted men. By July 3, 1864, at the time of the temporary dismounting of the regiment, the aggregate strength had dwindled to 433, a loss of 301 men, or about forty per cent. Sickness no doubt accounted for much of the drop, but constant battle from April 7 to July 3 must have accounted for at least one half of the loss. Only a few old newspapers have revealed about fifty casualities. Many, many more must have been killed wounded, or captured during the many hundreds of battles occurring after Pleasant Hill.[33] Several are known to have been casualties on May 13, 1864, when no fighting was reported to have taken place.

At one time the aggregate strength of the regiment was 892, but this included men of Mullin's company and Gray's company, who were transferred to Likens' regiment in the reorganization of the regiment in November, 1863.

COMPANY COMMANDERS

Company commanders on February 29, 1864, were as follows:

Company	Captain
A	William G. Hill
B	George W. Cooper
C	William P. Payne
D	George W. Diamond
E	Ruben A. Reeves
F	John K. Williams
G	Samuel H. B. Cundiff
H	James F. Warren
I	Paschal R. Turner
K	Russell J. Starr

FIELD AND STAFF

Field and staff members of the regiment on February 29, 1864, were as follows:

Regimental Duty—Name and Rank

Commander—Alexander Watkins Terrell, Colonel
Second in Command—John C. Robertson, Lt. Colonel
Third in Command and
  Battalion Commander—Hiram S. Morgan, Major
Battalion Commander—George K. Owens, Major

Adjutant—Robert M. Rutledge, 1st Lt.
Assistant Quartermaster—James E. Preston, Lt.
Instructor of Infantry Tactics—Charles E. Williamson, Lt.
Surgeon—Bennet L. Rye
Assistant Surgeon—Allen D. Harn
Assistant Surgeon—George Wyche
Sergeant Major—Francis M. Gaston
Ordnance Sergeant—Thomas A. Watson
Quartermaster Sergeant—Thomas T. Gammage
Commissary Sergeant—Addison Moyer
Chief Bugler—Martin V. Pierson
Assistant Commissary Sergeant
(possibly a lieutenant)—Theodore Lubbock

FOOTNOTES

[1]Microfilm 323, *Terrell's Texas Cavalry* Rolls 177, 178, 179.

[2]Elise Warenskjold, *The Lady With the Pen,* edited by L.A. Clausen, Norwegian-American Historical Association, Northfield, Minnesota, 1961, 99.

[3]*Terrell's Texas Cavalry Regiment,* microfilm in author's possession, original document on file in Washington, D.C., at National Archives.

[4]Microfilm 323, *Terrell's Texas Cavalry,* Rolls, 177, 178, 179.

[5]*Terrell's Texas Cavalry Regiment,* microfilm in author's possession, original document on file in Washington, D.C., at National Archives.

[6]Microfilm 323, *Terrell's Texas Cavalry,* Roll 177.

[7]*Terrell's Texas Cavalry Regiment,* microfilm in possession of the author prepared by the National Archives in Washington, D.C. Microfilm 323, *Terrell's Texas Cavalry,* Rolls 177, 178, 179.

[8]Microfilm 323, *Terrell's Texas Cavalry,* Rolls 177, 178, 179; *George Washington Diamond's Account of the Great Hanging at Gainesville, 1862,* ed. Sam Acheson and Julie Ann Hudson O'Connell, Austin, The Texas State Historical Association, 1963, 1.

[9]*Terrell's Texas Cavalry Regiment*, microfilm in possession of the author prepared by The National Archives in Washington, D.C.

[10]Microfilm 323, *Terrell's Texas Cavalry,* Rolls 177, 178, 179.

[11]Houston *Daily Telegraph*, June 8, 1864.

[12]Microfilm 323, *Terrell's Texas Cavalry,* Rolls, 177, 178, 179.

[13]*Terrell's Texas Cavalry Regiment,* microfilm in possession of the author prepared by the National Archives in Washington, D.C.

[14]The Galveston *Tri-Weekly News,* April 25, 1864.

[15]Houston *Daily Telegraph,* June 8, 1864.

[16]*Terrell's Texas Cavalry Regiment,* microfilm in possession of the author prepared by the National Archives in Washington, D.C.

[17]The Galveston *Tri-Weekly News,* April 25, 1864.

[18]Houston *Daily Telegraph,* June 8, 1864.

[19]Microfilm 323, *Terrell's Texas Cavalry,* Roll 177.

[20]*Terrell's Texas Cavalry Regiment,* microfilm in possession of the author prepared by the National Archives in Washington, D.C.

[21]The Galveston *Tri-Weekly News,* April 25, 1864.

[22]Zuber's father of Minden was a cousin to William Physick Zuber of Company H, 21st Texas Cavalry, which also saw duty in Louisiana. William Physick Zuber's memoirs were published in 1971. William Physick Zuber, *My Eighty Years in Texas,* edited by Janis Boyle Mayfield, Austin, University of Texas Press, 1971, 171.

[23]*Official Records,* VII, Series 2, 1092-1093.

[24]Microfilm 323, *Terrell's Texas Cavalry,* Roll 177.

[25]Microfilm 323, *Terrell's Texas Cavalry,* Rolls 177, 178, 179; *Terrell's Texas Cavalry Regiment,* microfilm in possession of the author prepared by the National Archives in Washington, D.C.

[26]The Galveston *Tri-Weekly News,* April 25, 1864.

[27]Microfilm 323, *Terrell's Texas Cavalry,* Roll 178.

[28]The Houston *Daily Telegraph,* June 8, 1864.

[29]The Galveston *Tri-Weekly News,* April 25, 1864.

[30]The Houston *Daily Telegraph,* June 8, 1864.

[31]The Houston *Daily Telegraph,* April 25, 1864.

[32]*Terrell's Texas Cavalry Regiment,* microfilm in possession of the author prepared by the National Archives in Washington, D.C.

[33]Winters, *The Civil War in Louisiana,* 428. In all, about 600 battles or skirmishes occurred in Louisiana during the Civil War, and most of these took place immediately preceding and after the battle at Pleasant Hill.

Appendix III

Roster Of The Regiment[1]

The roster was taken from the only known muster roll in existence for this regiment (see note No. 1), and was dated February 29, 1864. It was probably surrendered to the Federal troops at Shreveport after the war ended as this was where Terrell had taken his headquarters wagon when he found his regiment had disbanded at Wild Cat Bluff. The Trans-Mississippi Department headquarters was moved to Houston later, and the records could have been surrendered there. Regiments were required to submit muster rolls periodically, but those issued before that date which had been sent to headquarters were probably destroyed or lost. Those made subsequent to February 29, 1864, if any were made, were probably lost in the wild confusion immediately prior to and after the end of the war. Every effort has been made to copy the names exactly as they appear on the rolls. Sometimes this was extremely difficult because of the handwriting of the clerks who wrote the names.

There is evidence that Terrell attempted to obtain a copy of the muster rolls of his regiment. On February 12, 1908, he wrote the secretary of war in Washington, D.C., asking for a copy. The secretary wrote back asking why Mr. Terrell wanted the information, and his reply was that he was often asked to assist men to obtain refuge in state institutions, that he could not possibly remember all the men who had served him, and needed the muster rolls "in the name of humanity."[2]

There is no evidence that he ever got them.

TERRELL'S TEXAS CAVALRY REGIMENT

Field and Staff

Terrell, Alexander W., Colonel, appointed June 20, 1863. Age 37, 5' 9" tall, hazel eyes, dark complexion, dark hair. Civilian occupation, lawyer. Born in Virginia. Home, Austin, Texas.

Robertson, John C., Lt. Colonel, appointed June 20, 1863. Age 39, 5' 11" tall, black eyes, auburn hair. Civilian occupation, lawyer. Born in Hancock County, Georgia. Home, Tyler, Texas. Signed parole at Marshall, Texas, on July 31, 1865.

Morgan, Hiram S., Major, appointed at Galveston, Texas, on August 9, 1863. Home, Bastrop, Texas. Wounded severely in the left arm in the battle of Pleasant Hill, Louisiana on April 9, 1864.[3] Apparently was a lawyer in civilian life as he was appointed clerk of the Supreme Court of Texas on November 15, 1864. He resigned his commission in Terrell's regiment while the regiment was in camp in Winn Parish, Louisiana, December 6, 1864. The wound he received at Pleasant Hill had resulted in the loss of use of his left arm.

Owens, George K., Major, appointed June 12, 1863. Age 32, hazel eyes, dark complexion, dark hair. Civilian occupation, merchant. Born in Lewis County, Kentucky.

Rutledge, Robert M., First Lieutenant and Adjutant, appointed June 15, 1863. Transferred to Terrell's regiment from Griffin's Battalion of Texas Infantry on the same date. Was 5' 9'' tall, gray eyes, fair complexion, brown hair. Civilian occupation, apparently was a professional soldier. Born in Ireland. Resigned on July 19, 1864, because of the condition of his family. Asked for permission to join some military unit in Texas. Medical reason for resignation was dysentery.

Preston, James E., A.Q.M., probably lieutenant, appointed at Galveston, Texas, on June 13, 1863. Age 31, 5' 6'' tall, blue eyes, brown hair. Civilian occupation, clerk. Born Saratoga, New York.

Lubbock, Theodore, Assistant Commissary Sergeant, (possibly was a lieutenant) appointed June 13, 1863. Age 22, 5' 10'' tall, gray eyes, dark complexion, black hair, weight 150 pounds. Born Harris County, Texas. This was probably Theodore Uglow Lubbock, the adopted son of Governor Francis R. Lubbock, born December 24, 1841, in Houston. Young Lubbock first served in Terry's Texas Rangers (8th Texas Cavalry Regiment), attaining the rank of first sergeant in Company K, and was discharged on a certificate of disability on October 27, 1862, because of nearsightedness. Later it was apparently felt he could serve as a commissary and he joined Terrell's regiment when it was organizing. On his birthday in 1864 he married Miss Laura C. Files. He engaged in the furniture and general commission business in 1866 and in 1890 was elected to the Texas Senate from the ninth district.[4]

Bonner, J.I. Dr., Surgeon. Was appointed surgeon in Terrell's regiment after having served in Timmon's regiment as a private. Born November 6, 1828, in Claiborne, Alabama. Graduated from Univ. of Alabama in 1847. Came to Texas in 1853, where his father had come about three years earlier. He served with Terrell's regiment in Texas and Louisiana, but was compelled to resign in 1864 because of sickness, being brigade surgeon at the time. Resumed his medical practice a few years after the war in Fairfield, Texas.[5] He died February 19, 1900, and is buried in the cemetery at Eureka, Texas.[6]

Rye, Bennet L., Surgeon. Present at Galveston on February 29, 1864. No other information found.

Harn, Allen D., Assistant surgeon. Present at Galveston on February 29, 1864. No other information found.

Wyche, George, Assistant surgeon. Enlisted at Tyler, Texas, on February 4, 1863. Was also listed as a private in Company H on February 29, 1864, at Galveston, Texas.

Williamson, Charles E., 2nd Lieutenant, appointed at Galveston, Texas, on June 28, 1863. Was assigned to duty as instructor on infantry tactics.

Gammager, Thomas T., Master Sergeant, Quartermaster Sergeant, enlisted at Rusk, Texas, on May 26, 1863. Was with the horses on February 29, 1864. Formerly in Company D.

Watson, Thomas A., Sergeant, enlisted by J.E. Grey at Brenham, Texas, on July 1, 1863. Appointed Ordnance Sergeant from Company I on November 26, 1863. On the February 29, 1864, muster roll he is shown as being on detached service to Perry Landing on the Brazos River to get arms.

Gaston, Francis M., Sergeant-major, enlisted at Henderson, Texas, on April 20, 1863. Present on February 29, 1864, but had been listed as sick in November and December, 1863.

Moyer, Addison Q., Commissary Sergeant, enlisted at Nacogdoches, Texas, on April 27, 1863.

Pierson, Martin V., Chief Bugler, Age 41, 5' 10'' tall, hazel eyes, dark complexion, dark hair. Born Monroe County, Mississippi. Enlisted at Larissa, Texas, on July 1, 1863.

Clark, J.T., Major, Registered at the Provost Marshal's office on November 20, 1863 as being from Terrell's regiment. No other information found. Apparently this was a mistake on the part of the person who recorded the entry. His name is listed here because it is shown on the microfilm.

ROSTER OF TERRELL'S TEXAS CAVALRY REGIMENT

COMPANY A

CAPTAINS

Hill, William G., enlisted at Prairieville, Texas, on December 15, 1862. Age 38, 5' 10'', black eyes, dark complexion, black hair. Civilian occupation, farmer. Born Tuscaloosa, Alabama. Promoted to captain on December 20, 1863, when Captain Spikes resigned.

Spikes, Israel, elected December 15, 1862. Age 37, 6'4'', gray eyes, fair complexion, auburn hair. Civilian occupation, farmer. Born Clark, Alabama. Was sick on a surgeon's certificate in December, 1863. Resigned December 7, 1863, on a medical disability.

LIEUTENANTS

McCorquodale, John C., enlisted at Prairieville, Texas, on December 15, 1862. Age 45, blue eyes, fair complexion, black hair. Civilian occupation, farmer. Born Sampson, North Carolina. In November, 1863, he was sent in pursuit of deserters.

Harrison, Samuel M., enlisted at Prairieville, Texas, on December 15, 1862. Was appointed lieutenant on December 20, 1863. Age 43, 5' 9'', fair complexion, black hair. Born Tennessee. Resigned March 19, 1864.

Carlisle, Alexander E., enlisted at Prairieville, Texas, on December 15, 1862. No other information found.

SERGEANTS

Anderson, Burbon B., enlisted at Carthage, Texas, on August 7, 1863.

Kyser, John H., enlisted at Prairieville, Texas, on December 15, 1862. Was on detached service with the horses at Fayetteville, Texas.

Long, Alex, enlisted at Prairieville, Texas, on December 15, 1862.

Malone, William C., enlisted at Prairieville, Texas, on December 15, 1862.

Sewell, Jesse A., enlisted at Prairieville, Texas, on December 15, 1862.

Wilson, Thomas F., enlisted at Kaufman, Texas, on August 5, 1863.

CORPORALS

Anderson, Jeptha, enlisted at Carthage, Texas, on July 18, 1863.

Roach, Robert E., enlisted at Prairieville, Texas, on December 15, 1862. Present February 29, 1864.

Wilson, William M., enlisted at Prairieville, Texas, on December 15, 1862. Joined regiment from desertion at Camp Lubbock, Texas, on November 17, 1863. Wounded at the Battle of Mansfield or Pleasant Hill, Louisiana, on April 8 or April 9, 1864.[7] Name appears on a prisoner of war roll at Columbus, Texas, on June 26, 1865. Signed parole No. 33 on that date.

PRIVATES

Allen, David, enlisted at Prairieville on December 15, 1862. Applied for a pension in Oklahoma in 1915.

Britton, Sidney D., enlisted at Prairieville, Texas, on December 18, 1862.

Buchanan, George A., enlisted at Prairieville, Texas, on December 15, 1862.

Carlisle, John A., enlisted at Prairieville, Texas, on December 1, 1863. At Fayetteville with regimental horses.

Connell, John T., enlisted at Kaufman, Texas, on January 7, 1863.

Cotton, Weaver, enlisted at Prairieville, Texas, on December 15, 1862. At Fayetteville with horses. Joined from desertion November 7, 1863.

Cross, C.S., dropped as a deserter.

Dempsey, David, enlisted at Kaufman, Texas, on May 7, 1863. Joined by transfer. Cook in hospital.

Dickey, Thomas F., enlisted at Prairieville, Texas, on December 15, 1862.

Easterwood, Silas D., enlisted at Camp Scurry on June 7, 1863.

Elliott, John A., enlisted at Prairieville, Texas, on December 15, 1862. At Fayetteville with horses.

Flowers, John W., enlisted at Prairieville, Texas, on October 15, 1863. Wounded in battle of Mansfield or Pleasant Hill, La., April 8 or April 9, 1864.[8]

Foster, William M., enlisted at Prairieville, Texas, on December 15, 1862.

Fuller, Henry, enlisted at Prairieville, Texas, on December 15, 1862. With horses at Fayetteville, February 29, 1864.

Godfrey, Robert, enlisted at Canton, Texas, on April 23, 1863.

Goodnight, Henry, enlisted at Kaufman, Texas, October 15, 1863. Was later promoted to corporal. Born in Marshall County, Kentucky, on November 11, 1844. Goodnight was reared and educated in Henderson County, Texas. After the war he engaged in business and became a banker in Van Zandt County.[9]

Goodnight, Thomas, enlisted at Prairieville, Texas, October 15, 1863. At Fayetteville with regimental horses.

Graham, Wesley G., enlisted at Prairieville, Texas, December 15, 1862.

Graves, Thomas B., enlisted at Prairieville, Texas, December 15, 1862. Age 26, 5' 8'', gray eyes, dark complexion, black hair, farmer. Born Morgan County, Illinois. In November 1863 was on detached service with Lt. Col. Robertson.

Halliburton, F.M., enlisted at Camp White, Texas, March 28, 1863.

Ham, George W., enlisted at Camp Lubbock, Texas, November 16, 1863.

Heffington, John, enlisted at Prairieville, Texas, December 15, 1862.

Heffington, Thomas, enlisted at Prairieville, Texas, December 15, 1862. At Fayetteville with regimental horses.

Howes, John, no information on where and when enlisted.

Huff, Arthur D., enlisted at Prairieville, Texas, December 15, 1862. Present. Was on detached service with Co. B in Bell County in November, 1863.

Huff, Joseph, enlisted at Prairieville, Texas, December 15, 1862. Was on detached service with Co. B in November and December, 1863.

Huff, Thomas S., enlisted at Prairieville, Texas, December 15, 1862.

Jackson, Andrew, enlisted at Prairieville, Texas, December 15, 1862.

Johnson, Littleberry, enlisted at Canton, Texas, April 23, 1863.

Johnson, Matthew, enlisted at Prairieville, Texas, December 15, 1862. Left sick near Belton while on detached service with Co. B.

Johnson, William M., enlisted at Prairieville, Texas, December 15, 1862. Sick in hospital at Hempstead, Texas.

King, Edley M., enlisted at Prairieville, Texas, December 15, 1862.

Lide, James W., enlisted at Prairieville, Texas, December 15, 1862.

Long, Robert A.C., enlisted at Kaufman, Texas, August 5, 1863.

McKinney, Kincheon, enlisted at Prairieville, Texas, December 15, 1862. At Fayetteville with regimental horses.

McWilliams, John F., enlisted at Prairieville, Texas, December 15, 1862. At Fayetteville with horses. Was sick in the hospital at Chappel Hill in November, 1863.

Morris, Benjamin F., enlisted at Prairieville, Texas, December 15, 1862.

Norman, John, enlisted at Prairieville, Texas, December 15, 1862.

Northcut, Isaac T., enlisted at Camp White, Texas, March 28, 1862.

Perry, John, enlisted at Prairieville, Texas, December 15, 1862.

Pierce, James S., enlisted at Cedar Grove, Texas, August 3, 1863.

Pierce, Thomas, enlisted at Prairieville, Texas, October 15, 1863.

Pugh, John E., enlisted at Prairieville, Texas, December 15, 1862. Detailed as regimental butcher.

Rader, John H., enlisted at Kaufman, Texas, August 5, 1863. In hospital at Houston since November 23, 1863.

Reese, Thomas, enlisted at Brownsville, Texas, October 15, 1863.

Richardson, William, enlisted at Kaufman, Texas, May 7, 1863. Transferred to Co. A from Co. D in November, 1863. Detailed in Surgeon's Dept. driving ambulance.

Rogers, Samuel, enlisted at Kaufman, Texas, March 30, 1863.

Ross, Charles S., deserted at Camp Wharton December 31, 1863.

Rosson, Morgan C., enlisted at Prairieville, Texas, December 15, 1862. On daily duty as a pioneer.

Simpson, William, enlisted at Prairieville, Texas, December 15, 1862. On daily duty as a pioneer.

Sockwell, George N., enlisted at Prairieville, Texas, October 15, 1863.

Spencer, William, enlisted at Prairieville, Texas, December 15, 1862. Bugler.

Stephens, Clayton, enlisted at Prairieville, Texas, December 15, 1862.

Towles, John, enlisted at Prairieville, Texas, December 15, 1862. Present in February, 1864. Was with brigade ordnance clerk on the December 30, 1863, roll.

Turney, William, enlisted at Prairieville, Texas, December 15, 1862.

Ward, Jesse W., enlisted at Prairieville, Texas, December 15, 1862. At Fayetteville with horses.

White, Charles, enlisted at Greenville, Texas, July 10, 1863.

White, Thomas J., enlisted at Prairieville, Texas, December 15, 1862.

Williams, Asa, enlisted at Prairieville, Texas, December 15, 1862.

Williams, James O., enlisted at Prairieville, Texas, December 15, 1862. At Fayetteville with horses.

Zink, Eli, enlisted at Prairieville, Texas, December 15, 1862.

COMPANY B

CAPTAINS

Cooper, George W., elected at Greenville, Texas, on January 10, 1863. Age 38, 6' 0'' tall, blue eyes, light complexion. Civilian occupation, druggist. On February 29, 1864, he was on detached service to attend courts-martial at Camp Johnston. Resigned his commission January 10, 1865.

LIEUTENANTS

Hart, Prior, elected at Greenville, Texas, on January 10, 1863. Age 38, 6' 2'' tall, blue eyes, light complexion, light hair. Civilian occupation, farmer. Born Bedford County, Tennessee. In absence of Captain Cooper, Hart was commander of the company. There is some evidence that Prior Hart was promoted to captain. He took command of the company when Captain Cooper resigned, a position which warranted that rank. Also, a personal interview with Prior Hart's grandson, Mr. W.C. Hart of Greenville, indicated Hart to be a captain in command of the company. In the same interview, it was revealed that Prior Hart obtained the sword belonging to General Banks in the Red River Campaign and brought it home with him when the war ended.[10] Some of General Bee's cavalrymen were in Banks' headquarters by daylight on the day following the battle of Pleasant Hill; so it is entirely possible that General Banks in his haste to depart might have left his sword behind, and it was picked up by Hart.[11]

Bourn, Samuel, elected at Greenville, Texas, on January 10, 1863. Age 37, 6' 2'' tall, light complexion.

Horton, Peter J.V., elected at Greenville, Texas, on January 10, 1863. Age 36, 6' 2'' tall, blue eyes, fair complexion, light hair. Civilian occupation, farmer. Born Carroll County, Tennessee. Resigned January 10, 1865.

SERGEANTS

Mattox, Perry W., elected at Greenville, Texas, on January 10, 1863.
McBride, John, elected at Greenville, Texas, on January 10, 1863.
Moore, Napolean B., elected at Greenville, Texas, on January 10, 1863.
Norris, John B., elected at Greenville, Texas, on January 10, 1863.
Payne, William O.H., elected at Greenville, Texas, on January 10, 1863.

CORPORALS

Coleman, Thomas, elected at Greenville, Texas, on January 10, 1863.
Fuller, John M., elected at Greenville, Texas, on January 10, 1863.
Shook, John F., elected at Greenville, Texas, on January 10, 1863. At Fayetteville with the horses effective February 3, 1864.
Smith, Alexander F., joined by transfer at Greenville, Texas, on February 1, 1864.

PRIVATES

Arnold, Christopher E., enlisted at Greenville, Texas, January 12, 1863. Transferred to another unit February 1, 1864.
Arnold, Thomas J., enlisted at Greenville, Texas, January 10, 1863.
Arnold, William H., enlisted at Greenville, Texas, January 10, 1863.
Avdas, John, enlisted at Greenville, Texas, January 10, 1863. At Fayetteville with regimental horses.
Babb, Kibble T., enlisted at Greenville, Texas, June 1, 1863.
Barron, James W., enlisted at Greenville, Texas, January 10, 1863.
Bennett, George W., enlisted at Greenville, Texas, July 25, 1863.
Brittingham, Andrew M., enlisted at Greenville, Texas, January 10, 1863.
Brown, John S., enlisted at Greenville, Texas, June 1, 1863.
Brownlow, A., enlisted at Greenville, Texas, July 25, 1863. Transferred to Major Burnett's battalion in exchange for C.F. Smith.
Brumley, Benjamin F., enlisted at Greenville, Texas, January 10, 1863. At Fayetteville with regimental horses.
Caro, Thomas, enlisted at Greenville, Texas, November 28, 1863.
Carrell, N., Absent, sick.
Cornell, Nathan, enlisted at Greenville, Texas, January 10, 1863. Absent, sick since August 20, 1863.
Cox, James T., enlisted at Greenville, Texas, January 10, 1863. At Fayetteville with regimental horses.
Currie, William H., enlisted at Greenville, Texas, January 10, 1863.

Dawning, John R., enlisted at Greenville, Texas, December 25, 1863.

Dial, John C., enlisted at Greenville, Texas, January 10, 1863.

Dodd, Joseph, enlisted at Greenville, Texas, January 10, 1863. Signed parole No. 1545 at Columbus, Texas, August 2, 1865.

Dugan, George H., enlisted at Greenville, Texas, January 10, 1863. Age 28, 6' 0'' tall, dark eyes, dark complexion, dark hair. Born Montgomery County, Kentucky. Farmer. Discharged at Camp Groce, Texas, July 30, 1863, leg trouble.

Duke, John J., enlisted at Greenville, Texas, January 10, 1863. At Fayetteville with regimental horses.

Eakin, John, enlisted at Greenville, Texas, January 10, 1863.

Foster, Elkin, enlisted at Greenville, Texas, May 7, 1863.

Garrett, William, enlisted at Greenville, January 10, 1863.

Hamilton, E.S., Sick at horse camp at Fayetteville, Texas.

Hamilton, U.S., enlisted at Greenville, Texas, January 10, 1863. Sick at the horse camp at Fayetteville, Texas.

Harris, William A., enlisted at Greenville, Texas, January 10, 1863.

Harrison, Richard J., enlisted at Greenville, Texas, July 25, 1863.

Hobbs, Silas O., enlisted at Greenville, Texas, December 17, 1863.

Hogan, S., enlisted at Greenville, Texas, January 10, 1863. Absent, sick at Belton since January 1, 1864.

Hulse, Alex E., enlisted at Greenville, Texas, January 10, 1863. Was a wagon master in August and September of 1863. Wounded in Louisiana and later died of gangrene.[12]

Jackson, Robert, transferred to McMahan's artillery battery on February 3, 1864, per Gen. Order No. 33.

Johnson, Crayton L., enlisted at Greenville, Texas, July 25, 1863.

Kinney, J. Jackson, enlisted at Greenville, Texas, July 25, 1863. Absent, deserted since January 1, 1864.

Kitching, James W., enlisted at Greenville, Texas, June 1, 1863. Age 28, 5' 8'' tall, blue eyes, light complexion, light hair. Civilian occupation, saddler. Born Callaway County, Missouri. Furloughed from March 8, 1864, to April 8, 1864.

Knowles, William B., enlisted at Camp Groce, Texas, July 1, 1863. Substitute for M.H. Canselor.

Landrum, Mathew, enlisted at Greenville, Texas, January 10, 1863.

Long, William, (Name could be Lang), enlisted at Greenville, Texas, January 10, 1863. Age 38, 5' 11'', farmer in civilian life. Born Clairmont County Ohio. Was a teamster and ambulance driver. Discharged on a physical disability on August 26, 1864.

Lawson, Andrew, enlisted at Greenville, Texas, January 10, 1863.

Lee, Henry B., enlisted at Greenville, Texas, January 10, 1863. At Fayetteville with horses.

Lynch, John W., enlisted at Greenville, Texas, July 25, 1863.

Marshall, John L., enlisted at Greenville, Texas, January 10, 1863.

McBride, Daniel, enlisted at Greenville, Texas, January 10, 1863.

McDonald, Real, enlisted at Greenville, Texas, December 1, 1863.

McFadden, Samuel A., enlisted at Greenville, Texas, January 10, 1863. Absent, sick at Hempstead, Texas.

McKinney, Uriah, enlisted at Belton, Texas, September 27, 1863. At Fayetteville with horses since February 3, 1864.

McUlmurrey, James C., enlisted at Greenville, Texas, June 1, 1863.

Odell, Joel, enlisted at Greenville, Texas, January 10, 1863. Absent, sick since June 7, 1863.

Odell, Turner, enlisted at Greenville, Texas, January 10, 1863.

O'Neal, James M., enlisted at Greenville, Texas, January 10, 1863. Butcher.

Patterson, Robert S., enlisted at Greenville, Texas, January 10, 1863. With the horses at Fayetteville, Texas.

Payne, Wilson M., enlisted at Greenville, Texas, January 10, 1863. With the horses at Fayetteville, Texas.

Pine, Overton, enlisted at Greenville, Texas, June 1, 1863. With the horses at Fayetteville, Texas.

Pine, Rufus P., enlisted at Greenville, Texas, December 5, 1863.

Ramsey, John A., enlisted at Greenville, Texas, January 10, 1863. With the horses at Fayetteville, Texas.

Rattan, Thomas H., enlisted at Camp Dixie, Texas, February 1, 1864. Joined by transfer.

Renu, William C., enlisted at Greenville, Texas, June 1, 1863. On detached service attending the sick at the horse camp at Fayetteville, Texas. Was a hospital steward.

Requa, Austin C., enlisted at Greenville, Texas, January 10, 1863.

Roachell, Isam A., enlisted at Greenville, Texas, November 28, 1863.

Robertson, B.F., enlisted at Belton, Texas, December 8, 1863.

Shook, Thomas W., enlisted at Greenville, Texas, June 1, 1863. Absent, sick since November 21, 1863.

Smith, William D., enlisted at Greenville, Texas, February 1, 1863. The regimental return shows enlistment date as February 1, 1864.

Sullivan, Charles A., enlisted at Greenville, Texas, July 25, 1863.

Thompson, James A., enlisted November 28, 1863.

Turley, Archibald, enlisted at Greenville, Texas, January 10, 1863. With horses at Fayetteville, Texas.

Voiles, Rudolphus, enlisted at Greenville, Texas, January 10, 1863. With horses at Fayetteville, Texas as a teamster.

Voiles, Samuel, enlisted at Greenville, Texas, January 10, 1863.

Walker, Samuel B.W., enlisted at Greenville, Texas, February 1, 1864. Applied for an Oklahoma Confederate pension about 1915.

Wall, George W., enlisted at Greenville, Texas, July 25, 1863.

Wallace, Nathaniel, transferred to McMahan's battery of artillery on February 3, 1864.

White, Joseph, enlisted at Greenville, Texas, November 28, 1863. On detached service in Houston with Labor Bureau.

Whitehead, Allen N., enlisted at Wharton County, Texas, May 1, 1863. With horses at Fayetteville, Texas.

Whitman, Stephen C., enlisted at Greenville, Texas, January 10, 1863. Age 30, 6' 0'' tall, blue eyes, light complexion, auburn hair. Civilian occupation, farmer. Born Wythe County, Virginia. Was on detached service with the quartermaster. Was forage master.

Williams, John J., enlisted at Greenville, Texas, on January 10, 1863.

Winston, John L., enlisted at Greenville, Texas, on January 10, 1863.

COMPANY C

CAPTAINS

Payne, William Preston, enlisted at Waxahachie, Texas, on January 31, 1863. Age 39, 5' 10'' tall, gray eyes, light complexion, light hair. Civilian occupation, farmer. Born Franklin, Alabama.

LIEUTENANTS

Avriett, James A., enlisted at Athens, Texas, on January 31, 1863. Elected lieutenant March 7, 1863. Age 37, 5' 6'' tall, black eyes, dark complexion, black hair. Civilian occupation, merchant.

Ballow, John W., enlisted at Athens, Texas, on January 31, 1863. Age 42, 5' 11'' tall.

Reynolds, George W., enlisted at Athens, Texas, on January 31, 1863. Elected lieutenant March 7, 1863. Age 40, 5' 11'' tall, blue eyes, light complexion, dark hair. Civilian occupation, farmer. Born Oglethorpe County, Georgia.

SERGEANTS

Boyd, Larkin M., enlisted at Athens, Texas, on January 31, 1863. Sick in quarters on February 29, 1864.

Chandler, Alphonso, H., enlisted at Athens, Texas, on January 31, 1863.

Green, Ira, enlisted at Waxahachie, Texas, on February 7, 1863. On daily duty as a pioneer on February 29, 1864.

Roberts, William H., enlisted at Waxahachie, Texas, on March 22, 1863.

Sanders, William C., enlisted in Houston County, Texas, on March 4, 1863. Was appointed sergeant in place of J.M. Walston, who resigned February 1, 1864. Sanders saw some duty with Company B near Belton, Texas, in pursuit of deserters.

CORPORALS

Bradford, Powell H., enlisted at Athens, Texas, on January 31, 1863. Wounded severely in the leg in Louisiana on May 13, 1864.[13]

Chandler, John C., enlisted at Athens, Texas, on January 31, 1863.

McKnight, Samuel H.A., enlisted at Waxahachie, Texas, on March 12, 1863. February 3, 1864, was on detached service at Fayetteville, Texas, with the regimental horses.

Turner, Walter M., enlisted at Waxahachie, Texas, on February 7, 1863.

PRIVATES

Adair, John C., enlisted at Athens, Texas, on December 8, 1863. This is shown as a reenlistment. No information on previous service is indicated.

Alexander, Marcus L., enlisted at Waxahachie, Texas, on February 21, 1863. Absent. Sick at Camp Dixie since February 2, 1864. Was company bugler.

Andrews, John M., enlisted at Waxahachie, Texas, on January 31, 1863. Saw some service with Co. B near Belton.

Anderson, J.M., was assigned to duty in McMahan's artillery in February 1864.

Ard, Bradford H., enlisted at Athens, Texas, on January 31, 1863.

Avant, Willie W., enlisted at Athens, Texas, on January 31, 1863.

Brandy, Richard T., enlisted at Waxahachie, Texas, on February 7, 1863.

Brown, John P., enlisted at Athens, Texas, on January 31, 1863. Died at Camp Wharton of pneumonia on January 19, 1864.

Burks, Mason M., enlisted at Athens, Texas, on January 31, 1863.

Carew, W.L., Absent. Sick in hospital since November 26, 1863.

Carlile, James, enlisted at Athens, Texas, on January 31, 1863.

Carlile, John T., enlisted at Athens, Texas, on January 31, 1863.

Carver, John, enlisted at Athens, Texas, on September 1, 1863. With horses at the horse camp near Fayetteville, Texas.

Carver, Levi, enlisted at Athens, Texas, on January 31, 1863. Died at

Hempstead of dropsy December 20, 1863. Had $130.25 in personal effects at time of death.

Carver, William P., enlisted at Athens, Texas, on January 31, 1863. Absent. On detached service at the horse camp at Fayetteville, Texas.

Carver, John S., enlisted at Athens, Texas, on August 18, 1863.

Coleman, James M., enlisted at Waxahachie, Texas, on January 31, 1863.

Cowan, William D., enlisted at Athens, Texas, on January 31, 1863.

Davis, Aaron G., enlisted at Camp Farris, Texas, on February 21 1863.

Davis, Alfred W., enlisted at Athens, Texas, on January 31, 1863. Sick in hospital at Houston. Was assistant in blacksmith shop.

Davis, Robert A., enlisted at Waxahachie, Texas, on August 31, 1863. At horse camp at Fayetteville, Texas.

Douglas, William P., enlisted at Athens, Texas, on January 31, 1863. Served with Co. B on detached service near Belton for several months.

Etheridge, Caleb, enlisted at Athens, Texas, on January 31, 1863. Bugler.

Fulton, Ephraim W., enlisted at Athens, Texas, on January 31, 1863. Farrier. Was with the horses at Fayetteville on February 29, 1864.

Glaze, Joseph A., enlisted at Athens, Texas, on January 31, 1863.

Grant, James M., enlisted at Athens, Texas, on January 31, 1863.

Hart, William S., enlisted at Greenville, Texas, on January 10, 1863. Age 31, 6' 0'', gray eyes, light complexion. Civilian occupation, farmer. Born Bledsoe County, Tennessee. Received a medical discharge at Camp Groce, Texas, on July 30, 1863.

Hearn, James E., enlisted at Camp Lubbock, Texas, on December 10, 1863. Detached to go get a horse in December, 1863.

Hearne, Oren, enlisted at Waxahachie, Texas, on March 7, 1863. Absent. Detailed in the Ordnance Dept. in Houston since November 25, 1863.

Hogg, Holland, enlisted on January 31, 1863. Present.

Ingram, Charles W., enlisted at Athens, Texas, on January 31, 1863. Reported missing after the Battle of Mansfield or Pleasant Hill, La., April 8 or April 9, 1864.[14]

Johnson, Peter, enlisted at Waxahachie, Texas, on February 7, 1863. With horses at Fayetteville.

Jones, Martin, enlisted at Athens, Texas, on October 1, 1863. Died near Camp Dixie, Texas, of congestion on January 30, 1864.

Killen, John W., enlisted at Athens, Texas, on January 31, 1863.

Lemmon, Robert A., enlisted at Waxahachie, Texas, on February 7, 1863.

Lindsey, James T., enlisted at Athens, Texas, on December 3, 1863. Joined by enlistment while company was on the march to Matagorda, at camp on Navidad Creek.

Louis, A.W., sick in hospital at Houston.

Lusk, Samuel B., enlisted at Athens, Texas, on January 31, 1863. With horses at Fayetteville.

Malcolm, Samuel S., enlisted at Waxahachie, Texas, on March 11, 1863. Was on daily duty as a pioneer in December, 1863.

McFarlin, Benjamin P., enlisted at Waxahachie, Texas, on February 7, 1863.

McMillan, Johnson M., enlisted at Athens, Texas, on January 31, 1863. Left sick at Camp Dixie February 10, 1864.

Neal, Henry C., enlisted at Waxahachie, Texas, on July 1, 1863.

Nipp, Reuben, Went blind and deserted from the hospital at Columbus, November 10, 1863.

Nix, James, enlisted at Athens, Texas, on January 31, 1863.

Oliver, William A., enlisted at Athens, Texas, on March 23, 1863. Sick in the hospital at Houston. Was a cook.

Owen, Silas M., enlisted at Athens, Texas, on January 31, 1863.

Parks, Felix G., enlisted at Waxahachie, Texas, on February 7, 1863. Detailed in QM Dept. at Houston. Saddler.

Pate, Richard P., enlisted at Athens, Texas, on January 31, 1863.

Pearce, James, enlisted at Athens, Texas, on December 21, 1863. Reenlisted in company.

Powell, Elijah, enlisted at Athens, Texas, on January 31, 1863. With horses at Fayetteville, Texas.

Prince, William, enlisted at Waxahachie, Texas, on February 7, 1863.

Reece, John W., enlisted at Waxahachie, Texas, on February 7, 1863.

Reynolds, John, enlisted at Athens, Texas, on January 31, 1863. Absent. Sick in Henderson County on surgeon's certificate since April 16, 1863.

Slaughter, John W., enlisted at Athens, Texas, on January 31, 1863.

Smartt, J. Preston, enlisted at Waxahachie, Texas, on February 7, 1863. Served several months with Co. B on detached service near Belton.

Smith, William K., enlisted at Waxahachie, Texas, on March 5, 1863. Detailed in QM Dept. in Houston.

Spikes, Seborn S., enlisted at Camp Dixie, Texas, on January 28, 1864. Apparently transferred into the company from another unit.

Spikes, W., left sick in hospital at Virginia Point. Had chronic dysentery.

Thorn, John, transferred to McMahan's artillery battery in February, 1864.

Todd, William R., enlisted at Greenville, Texas, on January 10, 1863. Age 42, 6' 0'' tall, gray eyes, dark complexion, dark hair. Civilian occupation, shoemaker. Born Ashe County, North Carolina. Was assigned as a courier June, 1863. Discharged July 31, 1863, on disability. Had epilepsia caused by injury to skull when kicked by a horse.

Walling, Joseph D., enlisted at Athens, Texas, on January 31, 1863. Absent. Sick in Hill County on a surgeon's certificate.

Watson, James M., enlisted at Athens, Texas, on January 31, 1863. Resigned as sergeant February 1, 1864. Was a clerk and a teamster in Confederate army.

Wood, William R., enlisted at Athens, Texas, on January 31, 1863. Present, but sick in quarters.

COMPANY D

CAPTAINS

Diamond, George W., enlisted as a private in Company B, 3rd Texas Cavalry Regiment on May 7, 1861. Transferred to the 11th Texas Cavalry Regiment after the hangings at Gainesville, Texas, in 1862 and in the spring of 1863 joined Terrell's regiment with a company he had raised on the lower Brazos River.[15] Age 27, 6' 2'', fair complexion, light hair. Civilian occupation, lawyer and newspaper editor. Born De Kalb County, Georgia. (See Co. D sketches for additional information).

LIEUTENANTS

Cameron, James J., enlisted at Henderson, Texas, on April 1, 1863. Age 23, 5' 11'' tall. Born in Alabama. No other information available.

Crow, Milton M., enlisted at Henderson, Texas, on March 4, 1863. Age 36, 5' 6'', dark eyes, fair complexion, dark hair. Civilian occupation, farmer. Born Habersham County, Georgia.

SERGEANTS

Barnett, W.H., enlisted at Carthage, Texas, on April 22, 1863. On detached service with the horses at Fayetteville, Texas, on February 29, 1864.

Cuningham, J.H., enlisted at Henderson, Texas, on March 15, 1863. Killed in action in Louisiana on May 16, 1864, at the Battle of Mansura.[16]

Livsey, W.E., enlisted at Henderson, Texas, on April 11, 1863. At the time of the February 29, 1864, roll he was sick on a surgeon's certificate of disability. He signed parole No. 2 at Henderson, Texas, on July 12, 1865. He was an ordnance sergeant.

Osburn, S.H., enlisted at Carthage, Texas, on April 10, 1863.

Spivey, D.F., enlisted at Henderson, Texas, on April 1, 1863.

CORPORALS

Gilley, B.W., enlisted at Henderson, Texas, on April 1, 1863.

Lockridge, H.T., enlisted at Henderson, Texas, on March 18, 1863. On February 29, 1864 was on detached service with the horses at Fayetteville, Texas.

Stone, J.P., enlisted at Henderson, Texas, on April 11, 1863.

Walton, O.B., enlisted at Henderson, Texas, on May 1, 1863.

PRIVATES

Alexander, James A., turned over by the enrolling officer, but never reported.

Allison, Thomas, turned over by the enrolling officer, but never reported.

Armstrong, R.B., enlisted at Carthage, Texas, on April 16, 1863.

Arnold, W.J., enlisted at Henderson, Texas, on July 1, 1863.

Ash, J.F., enlisted at Henderson, Texas. Sick in field hospital at Hempstead since February 21, 1864.

Atkins, William, No information, other than being on detached service since June 1863.

Ballard, Joshua, No information, other than being on detached service as a teamster.

Barrentine, James, Absent. At the hospital at Virginia Point.

Berry, D.L., enlisted at Henderson, Texas, on July 10, 1863.

Bonland, W.A., enlisted at Shelbyville, Texas, on September 1, 1863. Was on the surrender roll at New Orleans, La., in May, 1865.

Branch, J.H., enlisted at Henderson, Texas, on July 15, 1863. Left sick in the hospital at Camp Dixie. Signed parole at Henderson, Texas, June 10, 1865.

Brown, George W., enlisted at Henderson, Texas, on March 20, 1863.

Bryan, E.F., enlisted at Carthage, Texas, on May 18, 1863.

Burns, W.P., enlisted at Shelbyville, Texas, on October 1, 1863.

Carter, J.W., Turned over by the enrolling officer, but never reported.

Cavin, J.B., enlisted at Henderson, Texas, on March 18, 1863. Sick in general hospital at Galveston. Signed parole at Henderson, Texas, July 12, 1865.

Childress, A.J., enlisted at Henderson, Texas, on March 10, 1863. Was on roll of prisoners of war at New Orleans May 26, 1865.

Clinton, T.M., enlisted at Henderson, Texas, on June 1, 1863. With the horses at Fayetteville, Texas.

Collins, A.B., enlisted at Henderson, Texas, on July 1, 1863. On extra duty in adjutant's office of regiment. Signed parole at Hempstead July 12, 1865.

Collins, M.W., enlisted at Henderson, Texas, on March 18, 1863.

Cooper, John, Turned over by the enrolling officer, but never reported.

Crow, Eli F., enlisted at Henderson, Texas, on March 28, 1863.

Cunningham, S.S., enlisted at Henderson, Texas, on December 16, 1863. Age 38, 6' 0" tall, blue eyes, dark complexion, dark hair. Was given a 60-day furlough from Natchitoches, La., April 21, 1865. Apparently was either wounded or had some sort of ailment. Later signed the prisoner of war roll at the end of the war June 12, 1865, at Henderson, Texas.

Delay, William, enlisted at Gilmer, Texas, on April 1, 1863. Age 32, 5' 6" tall, hazel eyes, dark complexion, dark hair. Civilian occupation, farmer. Born in Illinois. Discharged July 30, 1863, at Camp Groce, Texas, because of epilepsy.

Duth, John, Sick on surgeon's certificate.

Falwell, W.H., enlisted at Colorado City, Texas, on November 1, 1863.

Furgerson, John, enlisted at Henderson, Texas, on April 1, 1863. On extra duty as a teamster.

Griffin, Thomas, enlisted at Henderson, Texas, on August 1, 1863.

Ham, Giles, enlisted at Henderson, Texas, on July 1, 1863.

Hancock, G.B., Transferred to McMahan's battery February 3, 1864.

Harper, J.J., enlisted at Henderson, Texas, on July 1, 1863. Transferred to McMahan's battery of artillery February 3, 1864.

Harris, J.G., Turned over by the enrolling officer of Panola County, Texas, but never reported.

Henson, J.H., enlisted at Henderson, Texas, on May 18, 1863.

Herin, W.M., enlisted at Carthage, Texas, on April 16, 1863.

Higgins, W.T., enlisted at Camp Groce, Texas, on June 1, 1863.

Jackson, W.D., enlisted at Henderson, Texas, on June 1, 1863. With horses at Fayetteville.

Jimmerson, D.E., enlisted at Henderson, Texas, on July 1, 1863. With horses at Fayetteville.

Jimmerson, D.C., enlisted at Shelbyville, Texas, on October 10, 1863.

Jones, John E., enlisted at Henderson, Texas, on July 1, 1863. Absent. Sick in hospital at Hempstead. Paroled at Henderson, Texas, July 12, 1865.

Lawley, W., transferred to McMahan's battery of artillery on February 3, 1864.

March, A.M., enlisted at Henderson, Texas, on April 20, 1863.

Martin, W.H., enlisted at Henderson, Texas, on May 22, 1863. With horses at Fayetteville.

McClure, A.C., enlisted at Henderson, Texas, on October 12, 1863.

McLaughlin, P.H., turned over by the enrolling officer of Panola County, Texas, but never reported.

Mirgs, William, transferred to McMahan's battery of artillery on February 3, 1864.

Moore, John A., enlisted at Henderson, Texas, on March 4, 1863. On extra duty as regimental wagon master.

Morgan, S.F., enlisted at Henderson, Texas, on March 18, 1863. On extra duty at regimental headquarters.

Morris, David, enlisted at Henderson, Texas, on July 1, 1863.

Morris, J.L., enlisted at Henderson, Texas, on April 1, 1863.

Morrow, H.M., enlisted at Henderson, Texas, on July 1, 1863. Appears on prisoner of war roll at New Orleans, La., May 26, 1865.

Parker, John C., enlisted at Kaufman, Texas, on April 26, 1863. 5'6" tall, blue eyes. Civilian occupation, merchant. Born Mississippi. Discharged on June 30, 1863, Houston, Texas. Opacity of both eyes.

Penn, G.J., No other information except on June 25, 1864, he wrote the Federal commander in New Orleans asking permission to visit a sick friend in the hospital in New Orleans apparently near death.

Phillips, C.W., enlisted at Henderson, Texas, on April 1, 1863. With horses at Fayetteville.

Ramsey, Jap M., enlisted at Henderson, Texas, on April 16, 1863. With horses at Fayetteville.

Ray, Jones D., enlisted at Gilmer, Texas, on April 1, 1863. Age 42, 6'0", hazel eyes, dark complexion, dark hair. Born in Franklin County, North Carolina. Discharged at Camp Groce, Texas, July 17, 1863, because of chronic rheumatism.

Scott, William, enlisted at Henderson, Texas, on April 16, 1863.

Sheddan, A.A., enlisted at Henderson, Texas, on April 1, 1863. On March 5, 1864, was allowed to go home as his brother Joseph Shedden took his place. A.A. Shedden to return to the company as soon as Joseph became subject to conscription age on first day of May, 1864.

Shedden, Joseph, enlisted on March 5, 1864. Was not a member of the regiment, but took his brother's place while he returned to Henderson. Was not to serve longer than May 1, 1864.

Simmons, J.B., enlisted at Henderson, Texas, on April 9, 1863.

Snodgrass, B.J., enlisted at Henderson, Texas, on July 1, 1863.

Stiles, J.R., Transferred to McMahan's artillery battery February 3, 1864.

Stinston, Green, transferred to McMahan's artillery battery on February 3, 1864.

Stone, J.F., enlisted at Henderson, Texas, on July 1, 1863. With the horses at Fayetteville, Texas.

Strickland, John, enlisted at Henderson, Texas, on August 15, 1863. Joined by transfer on October 18, 1863.

Sullivan, John, enlisted at Henderson, Texas, on May 8, 1863. Was on a prisoner of war roll at Shreveport, La., sometime between May 26, 1865, and August 1, 1865.

Towns, A.L., transferred to McMahan's artillery battery on February 3, 1864.

Turnbow, James, enlisted at Henderson, Texas, on April 1, 1863. Left sick at Camp Sidney Johnston February 20, 1864. Was a teamster. Was also in Company H at one time.

Vincent, J.M., transferred to McMahan's artillery battery on February 3, 1864.

Whidden, W.A., enlisted at Henderson, Texas, on April 16, 1863.

Wilder, E.W., enlisted at Henderson, Texas, on April 10, 1863. With the horses at Fayetteville, Texas.

Williams, John, transferred to McMahan's artillery battery on February 3, 1864.

Yarbrough, Wiley, enlisted at Tyler, Texas, on February 4, 1863. Age 28, 5'9'' tall, blue eyes, fair complexion, auburn hair. Born St. Clair County, Alabama. Civilian occupation, physician. Discharged at Camp Groce, Texas, on July 21, 1863, because of a physical disability. Had received a knife wound in the lumbar region of the spine destroying the influence of the motary nerves of the left leg. This resulted in partial loss of use of the leg.

COMPANY E

CAPTAINS

Reeves, Reuben A., enlisted at Palestine, Texas, on April 11, 1863. Resigned September 28, 1864, by S.O. No. 264. He was elected associate justice of the Supreme Court of Texas in August of that year and resigned to take that position.

LIEUTENANTS

Gunnels, Nathan C., enlisted at Palestine, Texas, on April 11, 1863. Wounded, had arm broken in Battle of Mansfield or Pleasant Hill, Louisiana, on April 8th or April 9th 1864.[17]

Butler, Thomas M., enlisted at Palestine, Texas, on April 11, 1863.

Gresham, James N., enlisted at Palestine, Texas, on April 11, 1863. Was on detached service with the horses at Fayetteville, Texas, on February 20, 1864. No further record was found when he filed for a pension on June 1, 1916.

SERGEANTS

Clark, John M., enlisted at Palestine, Texas, on August 24, 1863.

Glenn, Richard C., enlisted at Palestine, Texas, on April 11, 1863. Was on a prisoner of war roll dated May 26, 1865.

Hassell, Spencer F., enlisted at Palestine, Texas, on July 11, 1863. Was on detached service with the horses at Fayetteville, Texas, on February 29, 1864.

McDonale, George H., enlisted at Palestine, Texas, on April 11, 1863.

McMillan, Zephaniah H., enlisted at Palestine, Texas, on June 5, 1863.

Teague, Drayton, enlisted at Palestine, Texas, on April 11, 1863.

CORPORALS

Gray, John, enlisted at Palestine, Texas, on July 25, 1863.

Madden, Robert J.S., enlisted at Palestine, Texas, on September 10, 1863.

McClanahan, William H., enlisted at Palestine, Texas, on June 5, 1863. Name appears on a prisoner of war roll in New Orleans, La., on May 26, 1865. (The middle initial could be an "M").

Moore, Thomas W., enlisted at Palestine, Texas, on June 5, 1863.

PRIVATES

Allen, W.J., enlisted at Palestine, Texas, on January 29, 1864.

Alverson, W.C., enlisted at Palestine, Texas, on August 1, 1864.

Barnett, Abner B., enlisted at Camp Wharton, Texas, on December 15, 1863.

Bartee, F.M., enlisted at Palestine, Texas, on April 11, 1863. Captured at Marksville, Louisiana, on May 16, 1864. Died May 27, 1864, of diarrhea. Buried in Cypress Grove, Grave No. 223.

Bowen, Wiley H., enlisted at Palestine, Texas, on April 11, 1863.

Butt, W.L., enlisted at Palestine, Texas, on April 11, 1863.

Calloway, W.J., enlisted at Palestine, Texas, on April 11, 1863. Absent, was detailed to work in a government shop on October 29, 1863.

Conniway, Alfred, enlisted at Palestine, Texas, on April 11, 1863. Appears on a prisoner of war roll at New Orleans May 26, 1865. A private Connery of Co. E was reported as being wounded in the Battle of Mansfield or Pleasant Hill on April 8th or 9th, 1864. Probably was this man.[18]

Conniway, James, enlisted at Palestine, Texas, on September 10, 1863.

Conwell, G.A., enlisted at Palestine, Texas, on February 9, 1864. Joined by enlistment.

Derden, R.W., enlisted at Palestine, Texas, on July 7, 1863.

Douglas, John, enlisted at Palestine, Texas, on September 10, 1863. Was on detached service with the horses at Fayetteville, Texas.

Dumas, J.K.M., enlisted at Palestine, Texas, on September 10, 1863. Died at Camp Dixie on January 28, 1864.

Derbin, Elam, enlisted at Palestine, Texas, on October 19, 1863.

Gibson, Thomas, Furloughed November 26, 1863, to go get a horse. No other information.

Gilmore, Bryant, enlisted at Palestine, Texas, on April 11, 1863.

Gilmore, R.H., enlisted at Palestine, Texas, on August 1, 1863. Was with the horses at Fayetteville, Texas, on February 29, 1864. Wounded slightly in the hand in the Battle of Mansfield or Pleasant Hill, Louisiana, April 8th or 9th, 1864.[19]

Glenn, N.A., enlisted at Palestine, Texas, on July 3, 1863. With the horses at the horse camp near Fayetteville, Texas, on February 29, 1864.

Graham, Jesse, enlisted at Palestine, Texas, on April 11, 1863.

Green, Samuel S., enlisted at Palestine, Texas, on September 11, 1863. Apparently signed a parole although there is no record of it in file.

Gwinn, William W., enlisted at Palestine, Texas, on April 11, 1863. Appears on a prisoner of war roll May 26, 1865.

Harecrow, W., enlisted at Palestine, Texas, on September 9, 1863.

Harrell, W.J., on detached service in the clothing store at Houston from February 15, 1864.

Haynes, M.H., enlisted at Palestine, Texas, on August 25, 1863. With the horses at the horse camp at Fayetteville, Texas, from February 15, 1864.

Heald, John M., enlisted at Palestine, Texas, on July 1, 1863. Age 25½, 5'5" tall, blue eyes, light complexion, dark hair. Civilian occupation, farmer. Born in Lawrence County, Mississippi. Discharged December 23, 1864, at the general hospital at Houston because of a fracture of the lower part of the tibia and other injuries. Injuries occurred November 26, 1863.

Helms, Cyrus, enlisted at Palestine, Texas, on April 11, 1863. Absent, sick at Hempstead, Texas, since January 1, 1864.

Hemby, John, enlisted at Palestine, Texas, on August 25, 1863. Died at Houston, Texas, on December 5, 1863, of German measles.

Herrell, W.J., enlisted at Palestine, Texas, on April 11, 1863. Transferred to the clothing bureau July 15, 1864.

Herrington, W.M., enlisted at Palestine, Texas, on April 11, 1863.

Hollmark, M.W., enlisted at Palestine, Texas, on April 11, 1863.

Hollmark, W.H., enlisted at Palestine, Texas, on April 11, 1863.

Holt, Rich, enlisted at Palestine, Texas, on September 12, 1863.

Hunter, John F., Was sick in the hospital at Houston in November, 1863. No other information.

Jones, G.T., enlisted at Palestine, Texas, on April 11, 1863.

Jones, Moses R., enlisted at Palestine, Texas, on August 1, 1863. Absent. Was detailed as a brigade teamster from December 22, 1863.

Joost, Alexander, enlisted at Palestine, Texas, on August 25, 1863. Was shown as a sergeant on a prisoner of war roll at New Orleans on May 26, 1865.

Langston, Willis B., Furloughed from Camp Lubbock, Texas, to go home for a horse and was not heard from again.

Laymance, A.J., enlisted at Palestine, Texas, on July 8, 1863.

Lawrance, J.R., Absent, was sick in the hospital at Houston.

Locker, R.H., enlisted at Palestine, Texas, on April 11, 1863.

Lyons, B.F., enlisted at Palestine, Texas, on August 13, 1863.

Maine, Elisha, enlisted at Camp Sidney S. Johnson on February 4, 1864. Was with the horses at Fayetteville, Texas, from February 4, 1864. Killed in the Battle of Mansfield or Pleasant Hill, Louisiana, April 8th or April 9th, 1864.[20]

Mane, Harman, enlisted at Palestine, Texas, on August 11, 1863.

McCallum, D.A., enlisted at Palestine, Texas, on August 28, 1863. With the horses at the horse camp near Fayetteville, Texas, from February 3, 1864.

McGill, John R., enlisted at Palestine, Texas, on April 11, 1863. Sick at the horse camp near Fayetteville, Texas, since February 3, 1864.

McKenzie, R.C., enlisted at Palestine, Texas, on August 20, 1863. With the horses at the horse camp near Fayetteville, Texas. Wounded in the left side on May 13, 1864, somewhere in Louisiana.[21]

McNeeley, George, enlisted at Palestine, Texas, on February 9, 1864. Joined by enlistment.

Mead, W.W., enlisted at Palestine, Texas, on September 11, 1863. With the horses at the horse camp near Fayetteville, Texas, from February 3, 1864.

Milam, W.B., enlisted at Palestine, Texas, on September 10, 1863.

Miller, John, enlisted at Palestine, Texas, on August 21, 1863.

Miller, T.B., enlisted at Palestine, Texas, on April 11, 1863. With the horses at the horse camp near Fayetteville, Texas, from February 3, 1864.

Milner, G.W., enlisted at Palestine, Texas, on August 25, 1863.

Morris, Samuel, enlisted at Palestine, Texas, on April 11, 1863. With the horses at the horse camp near Fayetteville, Texas, from February 3, 1864.

Owen, F.M., enlisted at Camp Dixie, Texas, on January 24, 1864.

Owens, Samuel T., enlisted at Palestine, Texas, on April 11, 1863. Wounded in the shoulder on May 13, 1864, somewhere in Louisiana.[22]

Page, James M., enlisted at Palestine, Texas on April 11, 1863.

Parker, John E., enlisted at Palestine, Texas, on September 10, 1863. With the horses at the horse camp near Fayetteville, Texas, from February 3, 1864.

Parker, N.D., enlisted at Palestine, Texas, on April 11, 1863. Absent. Sick at Hempstead, Texas, since February 15, 1864.

Pilgrim, Thomas, transferred to McMahan's artillery battery by S.O. No. 33 on February 3, 1864.

Pope, Hardy, enlisted at Palestine, Texas, on September 9, 1863. Age 42, 5'4'' tall. Civilian occupation, farmer. Born Marion County, Mississippi.

Posey, Adrian, enlisted at Palestine, Texas, on August 1, 1863. Transferred to the clothing bureau in Houston February 15, 1864. Name appears on a prisoner of war roll in New Orleans May 26, 1865. Posey was born on February 18, 1830, and died on January 7, 1913. He is buried in the Prairie Point Cemetery, south of Kerens, Texas.[23]

Powell, G.W., enlisted at Palestine, Texas on April 11, 1863.

Price, C.L., enlisted at Palestine, Texas, on August 25, 1863. Absent, sick at Hempstead, Texas, from January 10, 1864.

Rampey, D.N., enlisted at Palestine, Texas, on October 1, 1863.

Ratcliff, David A., enlisted at Palestine, Texas, on August 25, 1863. Signed parole No. 591 at Millican, Texas, on July 11, 1865. Name also appears on a prisoner of war roll at New Orleans May 26, 1865.

Redwine, Jacob J., enlisted at Palestine, Texas, on August 6, 1863. Absent, sick at Camp Dixie, Texas, since February 3, 1864.

Reeder, W., enlisted at Palestine, Texas, on January 20, 1864.

Reynolds, T.J., enlisted at Palestine, Texas, on April 11, 1863.

Rice, John T., enlisted at Palestine, Texas, on August 20, 1863. 5'10" tall, blue eyes. With the horses at the horse camp near Fayetteville, Texas, on February 29, 1864. Captured near Morganza, La., on August 8, 1864. Was sent to New York November 5, 1864. Took oath of allegiance to the U.S. at Elmira, N.Y., on June 19, 1865. Was released to return to his residence in Marshall, Texas, on the same date.

Ricks, T.L., enlisted at Palestine, Texas, on August 25, 1863. With the horses at the horse camp near Fayetteville, Texas, from February 3, 1864.

Saddler, N.F., enlisted at Palestine, Texas, on August 6, 1863.

Saunders, Lewis, enlisted at Palestine, Texas, on April 11, 1863.

Spikes, William, discharged on a surgeon's certificate November 23, 1863.

Stephens, J.F., enlisted at Camp Sidney F. Johnston, Texas, on February 4, 1864.

Stovall, John D., enlisted at Camp Wharton, Texas, on December 27, 1863. Name appears on a prisoner of war roll at New Orleans, La., on May 26, 1865. Shows his residence as Anderson County, Texas.

Vannoy, John H., enlisted at Palestine, Texas, on April 11, 1863.

Watts, John F., enlisted at Palestine, Texas, on April 11, 1863. Absent, sick on a surgeon's certificate from February 1, 1864. Name appears on a prisoner of war roll at New Orleans, La., on May 26, 1865.

Wood, John M., enlisted at Palestine, Texas, on April 11, 1863.

Woodard, Jonathan, enlisted at Palestine, Texas, on April 11, 1863. Absent, sick at the horse camp near Fayetteville, Texas, from February 3, 1864.

Woosley, David W., enlisted at Palestine, Texas, on August 13, 1863. Name appears on a prisoner of war roll at New Orleans, La., on May 26, 1865.

COMPANY F

CAPTAINS

Williams, John K., enlisted at Carthage, Texas, on April 1, 1863. Elected captain April 22, 1863.

LIEUTENANTS

Butler, William W., enlisted at Carthage, Texas, on April 16, 1863.
Ritter, Benjamin F., enlisted at Carthage, Texas, on April 20, 1863. Elected lieutenant April 25, 1863.

SERGEANTS

McMillin, Marshal, enlisted at Shelbyville, Texas, on August 5, 1863. Was also in Company A at one time.
Moore, James J., enlisted at Carthage, Texas, on April 16, 1863. With the horses at Fayetteville, Texas, on February 29, 1864.
Porter, James K., enlisted at Carthage, Texas, on April 16, 1863.
Soape, George G., enlisted at Carthage, Texas, on April 16, 1863.

CORPORALS

Davis, Henry L., enlisted at Carthage, Texas, on April 20, 1863.
Roggers, Nathanial A., enlisted at Carthage, Texas, on April 16, 1863. Was paid $150.80 on February 29, 1864, for service from September 1, 1863, through February 29, 1864.
Watson, John A., enlisted at Carthage, Texas, on May 20, 1863.
Weaver George N., enlisted at Carthage, Texas, on April 16, 1863.

PRIVATES

Barksdale, N.E., enlisted at Carthage, Texas, on October 1, 1863.
Barksdale, N.G., Was sick in November 1863.
Beaty, Francis M., enlisted at Shelbyville, Texas, on August 1, 1863. With the horses at the horse camp near Fayetteville, Texas.
Beaty, Leroy M., enlisted at Shelbyville, Texas, on August 1, 1863.
Bird, John W., enlisted at Carthage, Texas, on October 16, 1863. Absent, sick on a surgeon's certificate from January 15, 1864.
Bowlin, Solomon, enlisted at Carthage, Texas, on April 16, 1863.
Brown, Joel F., enlisted at Buena Vista (Pecos City), Texas, on April 20, 1863.
Burleson, B.D., Was sent home for a horse November 22, 1863. No other information shown.
Butler, Henry C., enlisted at Carthage, Texas, on April 16, 1863.

Casey, N.J., enlisted on February 20, 1864. Joined by enlistment.
Chandler, Allen, enlisted at Shelbyville, Texas, on July 20, 1863.
Chathma, James, enlisted at Carthage, Texas, on July 23, 1863.
Cooper, Joseph, enlisted at Carthage, Texas, on April 20, 1863.
Crawford, Thomas H., enlisted at Carthage, Texas, on April 20, 1863.
Cross, Albert M., enlisted at Carthage, Texas, on April 16, 1863.
Dillard, Andrew R., enlisted at Carthage, Texas, on April 20, 1863.
Doris, W.R., Transferred February 1864. No other information.
Easley, Newton J., enlisted at Columbia, Texas, on February 20, 1864.
Fenner, Benjamin R., enlisted at Columbus, Texas, on October 2, 1863.
Gilbreath, Thomas, enlisted at Carthage, Texas, on August 7, 1863.
Hall, Sidney E., enlisted at Carthage, Texas, on August 5, 1863. 5'10'' tall, dark eyes, fair complexion. Civilian occupation, farmer. Born Iredell County, N.C. Discharged December 13, 1863, at Camp Gulf Prairie, Texas, because of aneurism of the aorta.
Hartsfield, W.G., enlisted at Carthage, Texas, on October 1, 1863. Absent, sick.
Hawkins, John J., enlisted at Carthage, Texas, on April 16, 1863.
Holden, B.D., In November, 1863, was on detached service. No other information.
Holden, Benjamin F., enlisted at Columbus, Texas, on October 2, 1863. Absent, sick at the horse camp near Fayetteville, Texas.
Holland, Brook D., enlisted at Carthage, Texas, on August 7, 1863. Absent, sick on a surgeon's certificate from January 15, 1864.
Hopper, John, Transferred to McMahan's battery of artillery on February 3, 1864, at Camp Sidney Johnston.
Hughes, William P., enlisted at Buena Vista (Pecos City), Texas, on April 20, 1863. Absent, sick on a surgeon's certificate from January 25, 1864.
Johnson, Jeremiah, enlisted at Robertson City, Texas, on April 20, 1863.
Jolly, Alvis J., enlisted at Carthage, Texas, on April 16, 1863. Age 38, 5'11'' tall, blue eyes, fair complexion, light hair. Civilian occupation, farmer. Born North Carolina. Received a furlough from March 12, 1864, to April 12, 1864.
Joplin, Jesse A., enlisted at Camp Dixie, Texas, on January 27, 1864.
Keaton, John R., enlisted at Columbia, Texas, on February 20, 1864.
Lacy, Benjamin B., enlisted at Carthage, Texas, on August 7, 1863. Absent, sick at the Hempstead, Texas, hospital.
Latham, Archabald, enlisted at Carthage, Texas, on April 20, 1863. Was a blacksmith.
Lister, Sidney H., enlisted at Shelbyville, Texas, on September 1, 1863.

Luker, Isaac, enlisted at Camp Groce (near Hempstead, Tx.) on July 20, 1863. With the horses at the horse camp near Fayetteville, Texas, from February 3, 1864.

Martin, Toliver F., enlisted at Carthage, Texas, on May 20, 1863. Absent, sick at Hempstead, Texas, hospital. Was a blacksmith.

McCain, William L., enlisted at Carthage, Texas, on May 15, 1863. Was also in Company A at one time.

Miller, Henry, enlisted at Carthage, Texas, on August 7, 1863. Was also in Company A at one time.

Miller, L., Transferred to McMahan's battery of artillery on February 3, 1864.

Moore, Christopher V., enlisted at Columbia, Texas, on November 21, 1863.

Moorman, John H., Was sent home for a horse and apparently never returned.

Morris, Benjamin F., enlisted at Carthage, Texas, on May 12, 1863. With the horses near Fayetteville, Texas.

Murfree, S.L., transferred to McMahan's battery of artillery on February 3, 1864.

Nail, George, enlisted at Carthage, Texas, on August 4, 1863.

Parish, J.W., was apparently discharged on a disability from Camp Groce, Texas, probably sometime in 1863.

Parker, Redie, enlisted at Carthage, Texas, on April 16, 1863. With the horses near Fayetteville, Texas.

Paxton, William, enlisted at Carthage, Texas, on April 20, 1863.

Phillips, James, enlisted at Carthage, Texas, on August 7, 1863. With the horses near Fayetteville, Texas.

Pool, E.P., trasnferred to McMahan's battery of artillery on February 3, 1864.

Rainwater, W.W., transferred to McMahan's battery of artillery on February 3, 1864.

Read, William H., enlisted at Carthage, Texas, on April 20, 1863.

Russell, Phillip, enlisted at Columbia, Texas, on February 20, 1864.

Sinclair, James, enlisted at Columbus, Texas, on October 12, 1863.

Smith, Anderson, enlisted at Carthage, Texas, on May 15, 1863. Sick at the horse camp near Fayetteville, Texas.

Smith, Henry, enlisted at Carthage, Texas, on May 16, 1863.

Spivey, Elisha P., Sent home for a horse November 25, 1863, and did not return. (As of February 29, 1864).

Spivey, D.H., enlisted at Carthage, Texas, on April 16, 1863. Sick at Chappell Hill hospital.

Steel, William E., died of typhoid fever on November 6, 1863, at Columbus, Texas.

Strange, J.C., deserted May 16, 1863.

Sulivan, David M., enlisted at Carthage, Texas, on April 20, 1863.

Sweaza, Ellie L., enlisted at Carthage, Texas, on April 20, 1863.

Thomas, James, transferred to McMahan's battery of artillery on February 3, 1864.

Thomas, Joseph H., enlisted at Carthage, Texas, on May 20, 1863.

Upchurch, James M., enlisted at Carthage, Texas, on August 2, 1863.

Upchurch, Oliver P.H., enlisted at Carthage, Texas, on August 2, 1863. Teamster with the horses near Fayetteville, Texas.

Vaught, William H., enlisted at Carthage, Texas, on April 16, 1863.

Walton, Park M., enlisted at Carthage, Texas, on April 16, 1863. With the horses near Fayetteville, Texas.

Whitehead, Henry S., enlisted at Columbus, Texas, on October 3, 1863. With the horses near Fayetteville, Texas.

Whitehead, J., was on detached service in November 1863. No other information.

Wilkerson, James M., enlisted at Carthage, Texas, on April 16, 1863.

Willson, John W., enlisted at Columbus, Texas, on September 22, 1863.

Willson, William, enlisted at Carthage, Texas, on April 16, 1863.

COMPANY G

CAPTAINS

Cundiff, Samuel H.B., enlisted at Nacogdoches, Texas, on April 27, 1863. Age 26, 5'9'' tall, blue eyes, fair complexion, light hair. Civilian occupation, editor. Born Hampshire County, Virginia.

LIEUTENANTS

Doherty, John F.F., enlisted at Nacogdoches, Texas, on April 24, 1863. Elected lieutenant April 27, 1863. Age 43, 5'10'' tall, dark eyes, dark complexion, dark hair. Civilian occupation, farmer. Born Overton County, Tennessee. His name appears on a list of Confederate deserters who were sent to New Orleans from the Red River on April 21, 1864. Apparently he simply walked into the Federal lines and gave himself up. He took the oath of allegiance in New Orleans on April 25, 1864.

Grayson, Charles C., enlisted at Nacogdoches, Texas, on April 27, 1863. In December, 1863, he became sick on the way to join the regiment and was hospitalized in Hempstead, Texas. He resigned on January 18, 1864, on account of a disability. He was 5'11'' tall, blue eyes, fair complexion, dark hair. Civilian occupation was surveyor. Born Madison County, Alabama.

Grayson, G.S.H., enlisted at Nacogdoches, Texas, on April 27, 1863. Age 27, 5'11'' tall, blue eyes, fair complexion, light hair. Civilian occupation, farmer. Born Sabine Parish, Louisiana. Signed parole No. 291 at Marshall, Texas, on July 20, 1865.

Waller, Thomas J., enlisted at Nacogdoches, Texas, on April 24, 1863. In a letter on file he indicates that he was compelled to leave his home in Michigan because of his pro-Southern sentiments. He had a brother in Price's army and wanted to be reassigned to a regiment of Missouri cavalry. He resigned November 14, 1864, to be effective November 26, 1864. He was 30 years old at the time.

SERGEANTS

Davis, M.J., enlisted at Nacogdoches, Texas, on April 24, 1863. Was killed in action in the Battle of Mansfield or Pleasant Hill, Louisiana, April 8, 1864, or April 9, 1864.[24]

Martin, William, enlisted at Nacogdoches, Texas, on April 27, 1863. He was with the horses at the horse camp near Fayetteville, Texas, on February 29, 1864.

McKnight, D.M., enlisted at Nacogdoches, Texas, on April 27, 1863. Wounded slightly in the head on May 13, 1864, in Louisiana.[25]

Pike, Samuel W., enlisted at Nacogdoches, Texas, on April 24, 1863. Was on daily duty as issuing sergeant of the company.

CORPORALS

Allison, Peter, enlisted at Bastrop, Texas, on August 21, 1863. Absent, had been sick at Fayetteville, Texas, since February 3, 1864, at the time of this muster roll on February 29, 1864.

Cardenas, Olegarde, enlisted at Nacogdoches, Texas, on April 27, 1863.

Smith, B.M., enlisted at Nacogdoches, Texas, on April 27, 1863. Absent, was sick at the hospital in Hempstead, Texas, on February 29, 1864.

Stack, J.B., enlisted at Nacogdoches, Texas, on April 27, 1863.

PRIVATES

Alders, James S., enlisted at Nacogdoches, Texas, on April 27, 1863.

Allbright, J.D., enlisted at Columbus, Texas, on October 6, 1863. Deserted December 25, 1863.

Allbright, James, absent, sick with leave.

Allen, J., absent, furloughed.

Anderson, A.J., enlisted at Nacogdoches, Texas, on July 27, 1863. With the horses at the horse camp near Fayetteville, Texas.

Anderson, Littleton B., enlisted at Fayetteville, Texas, on October 6, 1863. Absent sick on a surgeon's certificate. Was on parole No. 1057 July 19, 1865, at Columbus, Texas.

Arero, Jose, with the horses at the horse camp near Fayetteville, Texas.

Bailey, William, enlisted at Nacogdoches, Texas, on April 27, 1863.

Barbo, Juan E., absent, sent home for a horse.

Baxter, Joseph W., enlisted at Nacogdoches, Texas, on April 27, 1863.

Bean, Edward J., enlisted at Bastrop, Texas, on November 14, 1863.

Bennett, John, enlisted at Bastrop, Texas, on August 21, 1863. With the horses at the horse camp near Fayetteville, Texas.

Best, Christopher, On November, 1863, he was listed as sick with the itch. No other information.

Blackburn, James G., enlisted at Nacogdoches, Texas, on April 24, 1863. With the horses near Fayetteville, Texas. He was a farrier.

Brimberry, W.A., enlisted at Nacogdoches, Texas, on April 7, 1863. With the horses near Fayetteville, Texas.

Brown, Elias, R., enlisted at Nacogdoches, Texas, on April 24, 1863. Absent, sick.

Chism, James A., enlisted at Nacogdoches, Texas, on April 4, 1863.

Clifton, Thomas B., enlisted at Nacogdoches, Texas, on April 27, 1863. Was on extra duty as a butcher.

Collins, Joshua, enlisted at Nacogdoches, Texas, on October 22, 1863.

Crawford, S.B., enlisted at Nacogdoches, Texas, on April 27, 1863.

Davidson, Oscar L., enlisted at Nacogdoches, Texas, on April 24, 1863. Captured by Federals near Morganza, La., September 11, 1864. Confined in Steam Levee Press No. 4, New Orleans. Transferred to Ship Island, Miss., October 5, 1864, and to New York November 5, 1864. Died of variola on December 12, 1864. Buried in grave No. 1154, Elmira N.Y.

Day, Eli A., enlisted at Nacogdoches, Texas, on April 24, 1863. Wounded in the leg May 13, 1864 in Louisiana.[26]

Dillard, John B., enlisted at Nacogdoches, Texas, on July 24, 1863. With the horses near Fayetteville, Texas.

Ebardo, Juan, enlisted at Nacogdoches, Texas, on April 24, 1863.

Fuller, T.H., enlisted at Nacogdoches, Texas, on July 24, 1863.

Furry, J.K.P., enlisted at Nacogdoches, Texas, on April 24, 1863.

Gaines, W.K., enlisted at Nacogdoches, Texas, on July 27, 1863.

Glass, W.S., enlisted at Nacogdoches, Texas, on April 27, 1863.

Hamilton, William H., no information shown on his enlistment.

Hardeman, Bunch, enlisted at Nacogdoches, Texas, on April 24, 1863. With the horses near Fayetteville, Texas.

Harwell, George W., enlisted at Nacogdoches, Texas, on April 24, 1863. Absent, sick in Nacogdoches.

Henson, Thomas J., enlisted at Nacogdoches, Texas, on April 27, 1863. Absent, sick in the hospital at Hempstead, Texas.

Hester, John R., enlisted at Nacogdoches, Texas, on April 24, 1863.

Hobbs, John R., enlisted at Nacogdoches, Texas, on April 24, 1863. Was a mail carrier in December, 1863.

Hodge, J.K.P., enlisted at Nacogdoches, Texas, on December 23, 1863.

Hyde, H.R., enlisted at Nacogdoches, Texas, on April 24, 1863.
Jennings, Thomas J., wounded at Battle of Mansfield or Pleasant Hill, La., April 8th or 9th, 1864.[27]
Johnson, J., in hospital at Columbus.
Johnson, R.H., enlisted at Nacogdoches, Texas, on April 27, 1863. Was a teamster in July, 1863.
Jones, A.F., enlisted at Nacogdoches, Texas, on October 22, 1863.
Keys, Thomas, enlisted at Nacogdoches, Texas, on April 27, 1863. With the horses near Fayetteville, Texas.
King, James M., enlisted at Nacogdoches, Texas, on April 24, 1863. With the horses near Fayetteville, Texas.
Long, F.M., enlisted at Nacogdoches, Texas, on April 24, 1863.
Mathews, H.J., enlisted at Nacogdoches, Texas, on October 22, 1863. With the horses near Fayetteville, Texas.
McClure, A.C., enlisted at Nacogdoches, Texas, on April 24, 1863.
McCormack, Edward, enlisted at Nacogdoches, Texas, on April 24, 1863. On extra duty as nurse in hospital.
McDaniel, W., enlisted at Nacogdoches, Texas, on April 24, 1863.
McKey, James D., enlisted at Nacogdoches, Texas, on October 22, 1863.
Melown, James, enlisted at Nacogdoches, Texas, on July 24, 1863.
Miller, J.H.M., enlisted at Nacogdoches, Texas, on July 24, 1863.
Mills, John, enlisted at Nacogdoches, Texas, on April 24, 1863. With the horses near Fayetteville, Texas.
Mims, W.P., enlisted at Nacogdoches, Texas.
Moore, William B., enlisted at Nacogdoches, Texas, on April 24, 1863. Wounded in the leg in Battle of Mansfield or Pleasant Hill, La., April 8th or 9th, 1864.[28]
Nearin, L.D., enlisted at Nacogdoches, Texas, on July 24, 1863. With the horses near Fayetteville, Texas.
Parrot, F.M., enlisted at Nacogdoches, Texas, on April 24, 1863.
Patterson, John, enlisted at Nacogdoches, Texas, on April 27, 1863.
Patterson, William, enlisted at Nacogdoches, Texas, on April 29, 1863.
Patton, R.C., enlisted at Nacogdoches, Texas, on July 27, 1863. On extra duty as regimental orderly.
Pike, Isaac N., enlisted at Nacogdoches, Texas, on November 14, 1863. Wounded in the arm and hip on May 15, 1864, in Louisiana.[29]
Pool, S.V., enlisted at Nacogdoches, Texas, on April 24, 1863. Absent with leave. May have had a substitute to serve for him.
Ray, John T.V., was on detached service with Lt. Col. Robertson at one time. No other information in file.
Robinson, RV., enlisted at Nacogdoches, Texas, on April 27, 1863. On detached service in the Q.M. Dept. in Houston.

Robinson, Samuel, enlisted at Nacogdoches, Texas, on January 1, 1864.

Rowe, W.W., enlisted at Nacogdoches, Texas, on April 27, 1863.

Rusk, John C., enlisted at Nacogdoches, Texas, on July 27, 1863. On extra duty as hospital steward.

Shepherd, W.J., enlisted at Nacogdoches, Texas, on April 27, 1863.

Sharley, W.W., enlisted at Nacogdoches, Texas, on July 27, 1863.

Smith, E.J., enlisted at Nacogdoches, Texas, on April 27, 1863.

Strode, Charles G., enlisted at Nacogdoches, Texas, on April 27, 1863. Was paid $144.80 for period September 1, 1863 to February 29, 1864.

Sutphen, A.J., enlisted at Nacogdoches, Texas, on April 24, 1863.

Swift, James, enlisted at Nacogdoches, Texas, on July 24, 1863.

Thomason, J.H.F., enlisted at Nacogdoches, Texas, on April 24, 1863. With the horses near Fayetteville, Texas.

Walker, James J., enlisted at Nacogdoches, Texas, on November 14, 1863.

Wallace, S.F., enlisted at Nacogdoches, Texas, on April 24, 1863. Age 39, 5'8" tall, gray eyes, dark complexion, dark hair. Civilian occupation, farmer. Was a teamster.

Ybarbo, Votal, enlisted at Nacogdoches, Texas, on April 27, 1863.

COMPANY H

CAPTAINS

Warren, James F., enlisted and elected captain at Tyler, Texas, on February 4, 1863. Age 30, 5'9½" tall, gray eyes, fair complexion, auburn hair. Civilian occupation, attorney. Born Henry County, Tennessee. Served about two years prior to service in Terrell's regiment with the Third Texas Cavalry as a private. He applied for a furlough on March 3, 1864, but probably did not get to take it as the regiment left shortly thereafter for Louisiana. He was in Company G at one time.

LIEUTENANTS

Garnett, James T., enlisted at Tyler, Texas, on February 4, 1863. Elected lieutenant November 14, 1863. Wounded in the knee in the Battle of Mansfield or Pleasant Hill, La., April 8th or 9th, 1864.[30] Signed parole Number 374 at Marshall, Texas, on July 1, 1865.

Gilley, Gabriel D., enlisted at Tyler, Texas, on February 4, 1863. Age 36, 6'2" tall, gray eyes, fair complexion, auburn hair. Civilian occupation, farmer. Born in Tennessee. In February 1864 was sick in the hospital at Hempstead, Texas. Was captured by the Federals near Morganza, La., on August 25, 1864. Was transferred for delivery and paroled on March 13, 1865.

Sigler, Byron C., enlisted at Tyler, Texas, on February 4, 1863. Age 36, 6'1'' tall, gray eyes, fair complexion, auburn hair. Born in Texas. On June 12, 1863, he purchased four yards of gray cloth for $24.00

SERGEANTS

Arnold, George M., enlisted at Tyler, Texas, on February 11, 1863.

Dobbs, William G., enlisted at Tyler, Texas, on February 14, 1863. Age 39, 5'11¾'' tall, gray eyes, fair complexion, auburn hair. Civilian occupation, farmer. Born Georgia.

Gillis, Daniel, enlisted at Tyler, Texas, on February 4, 1863. At the time of this muster roll on February 29, 1864, he was with the horses at the horse camp near Fayetteville, Texas.

Shelton, J.K.P., enlisted at Tyler, Texas, on February 24, 1863. Signed parole No. 445 at Marshall, Texas, on August 12, 1865.

Walker, Mason D., enlisted at Tyler, Texas, on February 4, 1863.

CORPORALS

Chetwood, John T., enlisted at Tyler, Texas, on March 19, 1863.

Dean, Alva W., enlisted at Tyler, Texas, on February 4, 1863. At the time of this muster roll he was with the horses at the horse camp near Fayetteville, Texas.

Dweggins, Daniel B., enlisted at Tyler, Texas, on February 4, 1863.

Stamps, Benjamin F., enlisted at Tyler, Texas, on February 4, 1863. Wounded in the shoulder in the Battle of Mansfield or Pleasant Hill, La., on April 8th or 9th, 1864.[31]

PRIVATES

Adams, John Q., enlisted at Tyler, Texas, on February 4, 1863. Absent, detached service in the Q.M. Dept. at Tyler.

Adams, Jeremiah, enlisted at Tyler, Texas, on February 4, 1863.

Allen, Young P., enlisted at Tyler, Texas, on February 4, 1863.

Altman, Nathan, enlisted at Tyler, Texas, on February 4, 1863.

Baron, James, enlisted at Tyler, Texas, on February 4, 1863. Died at Houston, Texas, on February 1, 1864 of typhoid fever.

Bateman, William B., enlisted at Tyler, Texas, on February 11, 1863. With the horses near Fayetteville, Texas.

Beason, E.G., enlisted at Tyler, Texas, on February 4, 1863. Absent, sick.

Boykin, William L., enlisted at Rusk Cty., Texas, on April 12, 1863. On extra duty as master clerk.

Boyles, Charles, enlisted at Tyler, Texas, on February 4, 1863.

Cain, Lewis J., enlisted at Tyler, Texas, on May 13, 1863.

Cane, Lewis D., absent, on detached service.

Carter, Edmond G., enlisted at Tyler, Texas, on February 4, 1863.

Cheatham, Charles P., enlisted at Tyler, Texas, on February 4, 1863.

Clark, James C., enlisted at Tyler, Texas, on June 15, 1863. Absent, sick in the hospital.

Clark, James E., Surgeon Rye found him unfit for duty because of chronic rheumatism. His name appears on a return of the post of Anderson, Grimes County, Texas, for August, 1864, as detailed by Capt. W. L. Goode, ordnance officer.

Coperland, J. G., enlisted at Tyler, Texas, on February 4, 1863.

Crabb, Thomas J., enlisted at Tyler, Texas, on February 4, 1863.

Dallehite, Jesse P., enlisted at Tyler, Texas, on April 19, 1863.

Davis, Anselum A., enlisted at Tyler, Texas, on February 24, 1863.

Day, Ballard A., enlisted at Tyler, Texas, on February 4 1863. 5'6½" tall, blue eyes, fair complexion, dark hair. Was with the horses near Fayetteville, Texas, on February 29, 1864. Captured near Morganza, La., on August 25, 1864. Signed oath of allegiance at Elmira N.Y., on July 7, 1865. Was one of the few of those captured who survived prison. He was released July 7, 1865.

Felton, James L., enlisted at Tyler, Texas, on April 17, 1863. Was age 24 as of August 15, 1864. Was a teamster in September, 1863. Appointed a corporal February 2, 1864. Was captured near Morganza, La., on August 25, 1864. Transferred to Ship Island, Miss., October 5, 1864, and sent to New York, November 5, 1864. He died April 20, 1865, of pneumonia and was buried in grave No. 1469 at Elmira, New York.

Fields, William, enlisted at Tyler, Texas, on April 10, 1863. Was sick in the hospital at Hempstead. His name appears on a "Register of Effects of Deceased Soldiers" which were turned over to the Q. M., C. S. A. The personal effects included $6.00 in cash.

Freday, Godfrey, enlisted at Tyler, Texas, on February 4, 1863.

Gentry, William, enlisted at Tyler, Texas, on February 4, 1863. Was a hospital steward.

Gilley, William, enlisted at Tyler, Texas, on October 7, 1863. With the horses near Fayetteville, Texas.

Goodman, Alex H., enlisted at Tyler, Texas, on February 4, 1863. Killed in action in the Battle of Mansfield or Pleasant Hill, La., April 8th or 9th 1864.[32]

Guthrey, William F., enlisted at Tyler, Texas, on August 15, 1863. With the horses near Fayetteville, Texas.

Hamilton, William F., enlisted at Tyler, Texas, on March 16, 1863.

Hampton, Elza, enlisted at Henderson, Texas, on April 1, 1863. Absent, sick in the hospital at Hempstead from January 15, 1864.

Hicks, John T., enlisted at Tyler, Texas, on February 4, 1863. At one time was detailed to wait on the sick in the hospital.

Hill, John P., enlisted at Tyler, Texas, on June 15, 1863.

Hodges, Harmon K., enlisted at Camp Dixie, Texas, on December 9, 1863.

Hodges, Jesse H., enlisted at Tyler, Texas, on February 4, 1863. He was born in Henderson County, Tennessee, February 2, 1826 and moved to Texas in 1848. After the war he returned to farming in Smith County and raised fourteen children.[33]

Howard, James L., enlisted at Tyler, Texas, on February 4, 1863.

Keys, John, enlisted at Tyler, Texas, on February 4, 1863. On detached service in the Ordnance Dept. at Marshall, Texas.

Kirkley, John, enlisted at Tyler, Texas, on May 19, 1863.

Knight, Thomas, enlisted at Tyler, Texas, on May 19, 1863.

Kuykendall, Eli, enlisted at Tyler, Texas, on February 3, 1864. Furloughed. Joined from the militia.

Larender, David S., enlisted at Tyler, Texas, on February 4, 1863.

Lote, Arthur J., enlisted at Tyler, Texas, on February 4, 1863.

Lott, William M., enlisted at Tyler, Texas, on August 1, 1863. With the horses near Fayetteville, Texas.

May, William H., enlisted at Shreveport, Louisiana, on May 8, 1863. Transferred to the 34th Texas Cavalry.

McDougal, James R., enlisted at Rusk, Texas, on April 15, 1863. Born March 4, 1838, in Alabama and came to Texas in 1856. Died July 9, 1905, at his residence near Tyler, Texas.[34]

McMurray, Anderson M., enlisted at Rusk, Texas, on April 18, 1863. Detailed as a pioneer in December, 1863.

Moore, Richard, enlisted at Tyler, Texas, on September 10, 1863.

Morris, James, enlisted at Tyler, Texas, on March 9, 1863.

Morrison, Eli B., enlisted at Shreveport, Louisiana, on August 5, 1863. Transferred to the 3rd Texas Cavalry.

Murphy, Archibald M., discharged December 31, 1863, because of a disability.

Murphy, L. B., absent, sick.

Murphy, Sidney B., enlisted at Rusk, Texas, on April 24, 1863.

Murphy, William, enlisted at Tyler, Texas, on February 11, 1863.

Norwood, James E., enlisted at Tyler, Texas, on February 4, 1863.

Norwood, J. L., sent for a horse in November, 1863. Absent without leave on the December, 1863, report.

Perry, Clark D., enlisted at Tyler, Texas, on February 4, 1863. He was born April 15, 1833, in Pickens Dist., S.C., moved to Texas in 1853. Perry died in Smith County, Texas, on November 29, 1899.[35]

Perry, William R., enlisted at Tyler, Texas, on February 4, 1863.

Praytor, Andrew J., enlisted at Tyler, Texas, on February 4, 1863. Wounded in the hip in the Battle of Mansfield or Pleasant Hill, La., on April 8th or 9th, 1864.[36]

Price, William W., enlisted at Tyler, Texas, on March 16, 1863. Absent, furloughed from the hospital at Columbus.

Prince, G.M., With the horses near Fayetteville, Texas.

Putty, Thomas J., enlisted at Tyler, Texas, on February 18, 1863.

Ray, John T., enlisted at Tyler, Texas, on February 4, 1863.

Reid, Stephen D., enlisted at Shreveport, Louisiana, on April 17, 1863.

Riddle, Thomas W., enlisted at Tyler, Texas, on February 4, 1863. In the convalescent hospital at Chappell Hill, Texas.

Rodgers, Hugh C., enlisted at Tyler, Texas, on March 16, 1863.

Rodgers, William C., enlisted at Tyler, Texas, on February 4, 1863. Sick in the hospital at Hempstead, Texas.

Rolin, Everett, enlisted at Tyler, Texas, on April 17, 1863. Sick at Camp Wharton, Texas.

Sears, Andrew B., enlisted at Tyler, Texas, on February 4, 1863. Age 46, 5'11'' tall, dark eyes, fair complexion, auburn hair. Civilian occupation, carpenter. Born in New York.

Sewell, J.M., enlisted at Tyler, Texas, on February 4, 1863.

Sheppard, William W., enlisted at Tyler, Texas, on March 19, 1863. With the horses near Fayetteville, Texas.

Sigler, Rufus B., enlisted at Denton, Texas, on May 1, 1863. Age 36, 6'0'' tall, dark eyes, light complexion, dark hair. Civilian occupation, lawyer. Born Perry County, Alabama. Had a wife and four children in Tyler, Texas. On February 29, 1864, he was on extra duty with the Quartermaster Department. Apparently enlisted initially on December 20, 1862.

Smith, Daniel W., enlisted at Tyler, Texas, on February 4, 1863. With the horses near Fayetteville, Texas.

Spruce, George W., enlisted at Tyler, Texas, on February 4, 1863. With the horses near Fayetteville, Texas.

Stone, John T., enlisted at Rusk, Texas, on June 15, 1863.

Swan, Thomas C., enlisted at Cherokee County, Texas, on March 13, 1863. Was sick at a private house in November and December of 1863, was present in Galveston on February 29, 1864.

Weaver, Jeremiah, enlisted at Tyler, Texas, on February 4, 1863.

Weaver, Joshua H., enlisted at Tyler, Texas, on May 14, 1863. With the horses near Fayetteville, Texas.

White, Robert F., enlisted at Tyler, Texas, on February 4, 1863.

Whitehorn, James E., enlisted at Tyler, Texas, on February 4, 1863. Absent, on furlough.

Wyche, George, enlisted at Tyler, Texas, on February 4, 1863.

Zuber, William, enlisted at Tyler, Texas, on February 4, 1863. Age 18, single. Born in Georgia. Captured near Morganza, La., on August 25, 1864. Was still at New Orleans on August 31, 1864. Was at Steam Levee Press No. 4 in New Orleans on September 28, 1864. He was transferred to

Ship Island, Mississippi, on October 5, 1864. Was transferred to New York on November 5, 1864, and arrived at Fort Columbus, New York Harbor, on November 16, 1864. Died February 26, 1865, at Elmira, New York, of chronic diarrhea. He was prisoner of war No. 1918 and was buried in grave No. 2379 at Elmira, New York.

COMPANY I

CAPTAINS

Murray, Caldean G., enlisted April 27, 1863. Age 33, 6'1'' tall, blue eyes, dark complexion, black hair. Civilian occupation, farmer. Born Jackson, Alabama. Deserted with about 100 men in September, 1863, when ordered to dismount and go to Galveston. Was in the hospital of the 13th Texas Infantry on December 17, 1863. Was sent to Houston, January 16, 1864.

Taylor, H.J., enlisted at Camp Lubbock, Texas, on November 11, 1863. Resigned November 4, 1863. Had a disability from a fall from a horse.

Turner, Paschal R., enlisted at Bastrop, Texas, on August 21, 1863. Was on detached service with the horses at the horse camp near Fayetteville at the time of this muster roll on February 29, 1864.

LIEUTENANTS

Burleson, Jacob, enlisted at Bastrop, Texas, on August 21, 1863.

Click, H.C., resigned December, 1863, on a disability. No other information.

Henry, Wood D., enlisted April 27, 1863. Age 47, 5'8'' tall, black eyes, dark complexion, black hair. Civilian occupation, farmer. Born Rhea County, Tennessee.

Moncure, John J., enlisted at Bastrop, Texas, on August 21, 1863.

Persons, John, was on detached service with Lt. Col. Robertson at one time. Died at Tyler, Texas, on November 1, 1863, while the regiment was at Camp Wharton, Texas.

SERGEANTS

Burgess, Gideon, enlisted at Bastrop, Texas, on August 21, 1863. Resigned November 9, 1864.

Burleson, Cumins, enlisted at Bastrop, Texas, on August 27, 1863.

Clanton, James H., enlisted at Bastrop, Texas, on August 21, 1863.

King, F.A., enlisted at Bastrop, Texas, on August 21, 1863. Age 37, 6'0'' tall, blue eyes, light complexion, dark hair. Civilian occupation, farmer. Born Wayne County, Tennessee.

Olive, James, enlisted at Bastrop, Texas, on August 21, 1863. Was also in Mullin's Company at one time.

Reding, James M., enlisted at Bastrop, Texas, on August 21, 1863. Wounded in the arm at the Battle of Mansfield or Pleasant Hill, La., April 8th or 9th, 1864.[37]

Sims, J.C., enlisted at Bastrop, Texas, on August 21, 1863.

CORPORALS

Cope, David, enlisted at Bastrop, Texas, on August 21, 1863. (Could be Coup).

Mays, D., No further information except that he was on detached service in November, 1863.

Mays, Samuel J., enlisted at Bastrop, Texas, on August 21, 1863.

Reager, George, enlisted at Bastrop, Texas, on August 21, 1863. Age 25, 6'0'' tall, blue eyes, light complexion. Born in Logan County, probably Kentucky. Captured near Morganza, La., on August 25, 1864. Was in the St. Louis General Hospital in New Orleans, La., on September 1, 1864, under treatment. Was transferred to Ship Island, Mississippi, October 5, 1864, and on to New York on November 5, 1864. He died in New York in November, 1864, of diarrhea.

Yancy, Napoleon B., enlisted November 14, 1863, at Columbus. He signed a prisoner of war parole at Columbus, Texas, on July 12, 1865.

PRIVATES

Baker, L.D., enlisted at Bastrop, Texas, on August 21, 1863.

Ballard, W.M.C., Was paroled August 31, 1865. Residence was shown as Rusk County, Texas.

Barbee, K.H., age 41, dark eyes, fair complexion, dark hair, 5'11'' tall. Captured near Morganza, La., on August 25, 1864. He was in the hospital with bronchitis on February 23, 1865, at Wayside Hospital in Richmond, Va. Was furloughed on February 24, 1865, from the general hospital at Howard's Grove, Richmond, Virginia.

Batcher, Fred, (or Butcher), enlisted at Columbus, Texas, on October 23, 1863. Was on detached service at Columbus, Texas, at the time of the February 29, 1864, muster roll.

Bennett, William, enlisted at Bastrop, Texas, on December 15, 1863.

Black, John, enlisted at Bastrop, Texas, on August 21, 1863.

Blaylock, Jef, enlisted at Bastrop, Texas, on August 21, 1863.

Buchanan, Mike, enlisted at Bastrop, Texas, on December 15, 1863. With the horses near Fayetteville, Texas.

Conway, R.M., enlisted at Bastrop, Texas, on August 21, 1863. On February 16, 1864, was sent on detached service to raise a company of men for Col. Shilton's regiment.

Cruiser, A.D., was on detached service in Colorado County.

Delmar, Henry E., enlisted at Camp Turner, Texas, on October 8, 1863.

Denson, John M., enlisted at Bastrop, Texas, on April 21, 1863. Was in the hospital at Shreveport June 30, 1864, to July 6, 1864. Probably was wounded in battles of Mansfield or Pleasant Hill. Was later transferred to the hospital in Keatchie, Louisiana.

Dickens, A.J., enlisted at Bastrop, Texas, on August 21, 1863. Deserted at Camp Wharton on January 10, 1864.

Frey, King L., enlisted at Bastrop, Texas, on August 21, 1863. With the horses near Fayetteville, Texas.

Furgison, John L., enlisted at Bastrop, Texas, on August 21, 1863. In September, 1863, he was transferred to Company H of Col. Cook's regiment because he had no horse.

Garrett, S.G., enlisted at Bastrop, Texas, on August 21, 1863. With the horses near Fayetteville, Texas.

Hamilton, N.P.S., enlisted at Bastrop, Texas, on December 13, 1863.

Hill, John, enlisted at Bastrop, Texas, on December 15, 1863.

Hopkins, J.F., enlisted at Bastrop, Texas, on December 15, 1863. Sick at Fayetteville.

James, R.F., Discharged at Camp Groce. No other information.

Jones, H.W., enlisted at Bastrop, Texas, on December 15, 1863. Wounded in the foot at the Battle of Mansfield or Pleasant Hill, La., on April 8th or 9th, 1864.[38] Was still in the hospital at Keatchie, Louisiana, on April 17, 1864.[39]

Kutchiman, K. (or Kutchgammas), enlisted at Columbus, Texas, on October 28, 1863.

Lacruise, Jo, enlisted at Columbus, Texas, on October 28, 1863.

Lawrence, George, enlisted at Bastrop, Texas, on August 21, 1863.

Lawrence, J.L., Was on detached service in Colorado County. No other information.

Mayblum, Nathan, enlisted at Columbus, Texas, on December 1, 1863. Absent, sick at the Columbus hospital.

Mitchell, Fielding F., Age 34, 5'10'' tall, blue eyes, black hair. Civilian occupation, farmer. Born in Mississippi. Residence was in Wood County, Texas. Received a medical discharge at Camp Martin on September 9, 1864.

Morgan, T.D., enlisted at Bastrop, Texas, on August 21, 1863. With the horses near Fayetteville, Texas.

Munt, Louis, enlisted at Columbus, Texas, on October 28, 1863.

Newsom, Edward, enlisted at Columbus, Texas, on November 24, 1863. Died of congestion of the brain at Camp Sidney Johnston, February 8, 1864.

Ogden, William, enlisted at Bastrop, Texas, on August 21, 1863.

Oliver, J.R., enlisted at Bastrop, Texas, on August 21, 1863.

Parkison, M.M. enlisted at Bastrop, Texas, on August 21, 1863. Was on extra duty to repair the wagons at the time of the February 29, 1864, muster roll.

Parr, Henry, enlisted at Columbus, Texas, on October 28, 1863. Transferred to the 5th Texas Cavalry on February 25, 1864.

Peirce, Thomas (Probably should be Pierce) No information except that he was on detached service with Lt. Col. Robertson at one time.

Phillips, Cornelius, enlisted at Bastrop, Texas, on August 21, 1863. Age 60, 6'4'' tall, blue eyes, gray hair. Civilian occupation, carpenter. Discharged because of age on September 30, 1864.

Potter, T.J., enlisted at Bastrop, Texas, on August 1, 1863. Deserted at Bastrop on December 25, 1863.

Rhodes, James, enlisted at Bastrop, Texas, on December 15, 1863.

Rhodes, Levin, enlisted at Bastrop, Texas, on December 15, 1863.

Rhodes, William, enlisted at Bastrop, Texas, on August 21, 1863.

Rice, John D., enlisted at Bastrop, Texas, on August 21, 1863. With the horses near Fayetteville, Texas.

Robertson, J.C., enlisted at Columbus, Texas, on December 4, 1863.

Shiller, Samuel, enlisted at Columbus, Texas, on November 6, 1863.

Snelling, John O., enlisted at Bastrop, Texas, on August 21, 1863. Was on extra duty to repair the wagons at the time of the February 29, 1864, muster roll. Probably near Fayetteville.

Spence, John D., enlisted at Bastrop, Texas, on August 21, 1863. With the horses near Fayetteville, Texas.

Stewart, Isaac R., enlisted at Camp Turner, Texas, on October 6, 1863. Paroled at Columbus, Texas, on July 19, 1865. Parole No. 1056.

Taylor, A.J., enlisted at Camp Turner on October 20, 1863. Absent, sick in the hospital at Columbus, Texas.

Tucker, G.W., enlisted at Bastrop, Texas, on December 5, 1863.

Waddle, D.M., enlisted at Bastrop, Texas, on December 15, 1863.

Waddle, Ellison, enlisted at Bastrop, Texas, on August 21, 1863. With the horses near Fayetteville, Texas, on February 29, 1864. Was wounded in the battle of Mansfield or Pleasant Hill, Louisiana, on April 8th or 9th, 1864.[40] Was still in the hospital (Eagan) in Keatchie, La. on April 17, 1864.[41]

Waddle, Nat, enlisted at Bastrop, Texas, on December 15, 1863.

Wilson, James, enlisted at Bastrop, Texas, on August 21, 1863.

Wilson, J.M., No information, except that he was on extra duty with the enrolling officer in Colorado County on November 30, 1863, and was sick in Columbus in December, 1863.

COMPANY K

CAPTAINS

Starr, Russell J., enlisted in Smith County, Texas, on April 6, 1863. Elected lieutenant April 27, 1863. Promotion date to captain not shown. Age 31, 5'8'' tall, blue eyes, fair complexion, light hair. Civilian occupation, farmer. Born Henry County, Georgia. Wounded in the hand about May 14, 1864, in Louisiana. Signed parole No. 305 at Marshall, Texas, on July 22, 1865.

LIEUTENANTS

Chancellor, Jessie G., enlisted April, 1863. Age 38, 6'0'' tall, blue eyes, fair complexion, light hair. Civilian occupation, merchant. Born in Alabama. In September, 1863, he and Capt. Murray led about 100 men in a mass desertion from the regiment. His name appears on a list of prisoners confined in the guardhouse in Galveston, Texas, on June 28, 1864. He signed parole No. 427 at Marshall, Texas, on August 10, 1865. He signed as Company D, 2nd lieutenant.

Dye, James P., enlisted at Cherokee County, Texas, on February 20, 1863. Was detailed as the bearer of dispatches by Lt. Col. Robertson in February, 1864.

Kincheloe, Elijah B., appointed lieutenant on April 1, 1863. Age 35, 5'11'' tall, blue eyes, fair complexion, dark hair. Civilian occupation, farmer. Born Hawkins, Tennessee. Resigned at Camp Groce, Texas (near Hempstead), on July 18, 1863.

Myers, David H., enlisted in Cherokee County, Texas, on February 10, 1863. Resigned March 19, 1864. No other information shown in the records.

Spradling, Robert W., enlisted in Wood County, Texas, on April 3, 1863. Age 37, 5'9'' tall, blue eyes, fair complexion, dark hair. Civilian occupation, farmer. Born Jefferson, Alabama. Wounded in the left side May 13, 1864, in Louisiana.[42] Was still in the hospital in Keatchie, Louisiana, on April 17, 1864.[43]

Trowell, John H., elected lieutenant April 1, 1863. Age 34, 6'2'' tall, dark eyes, dark complexion, dark hair. Civilian occupation, editor. Born Wilcox, Alabama. Residence Upshur County, Texas. Signed parole No. 307 on July 22, 1865, at Marshall, Texas.

Dublin, A.C., enlisted in Cherokee County, Texas, on February 10, 1863.

Morgan, Francis H., enlisted in Cherokee County, Texas, on February 10, 1863.

Snow, C.C., enlisted in Cherokee County, Texas, on August 25, 1863.

Weisner, W.O., enlisted in Cherokee County, Texas, on February 10,

1863. Killed in Action in the battle of Mansfield or Pleasant Hill, Louisiana, April 8th or 9th, 1864.[44]

Yarberry, T.T., enlisted in Cherokee County, Texas, on February 10, 1863.

CORPORALS

Beaird, Benjamin W., enlisted in Cherokee County, Texas, on February 10, 1863. Was sick in the hospital in Hempstead, Texas, at the time of the February 29, 1864, muster roll.

Kirkland, James M., enlisted in Cherokee County, Texas, on February 10, 1863.

Porter, John H., enlisted in Cherokee County, Texas, on February 10, 1863. Dropped from the roll as a deserter February 1, 1864. He had been sent for a horse on November 25, 1863, and did not return.

Robins, Benjamin L., enlisted in Wood County, Texas, on April 3, 1863. Age 38, 5'10'' tall, blue eyes, fair complexion, light hair. Civilian occupation, farmer.

Scott, Samuel L., enlisted in Cherokee County, Texas, on February 10, 1863. Wounded in the battle of Mansfield or Pleasant Hill, La., on April 8th or 9th, 1864.[45]

PRIVATES

Anderson, Wade H., enlisted in Smith County, Texas, on September 28, 1863.

Beard, John T., enlisted in Cherokee County, Texas, on August 8, 1863. Was sick in the hospital in Hempstead, Texas, at the time of the February 29, 1864, muster roll.

Bell, C.C., absent without leave at the time of the February 29, 1864, muster roll.

Bell, Richard S., enlisted in Cherokee County, Texas, on February 10, 1864.

Berry, C.J., enlisted at Camp Wharton, Texas, on December 20, 1863.

Berry, Elisha F., enlisted in Van Zandt County, Texas, on April 22, 1863.

Blantern, Charles, enlisted in Cherokee County, Texas, on August 25, 1863.

Bobbitt, James L., enlisted in Cherokee County, Texas, on August 8, 1863.

Bobbitt, John B., enlisted in Cherokee County, Texas, on July 1, 1863.

Brown, Nelson, enlisted in Wood County, Texas, on April 3, 1863.

Broyles, Benjamin A., enlisted in Cherokee County, Texas, on July 1,

1863. Cherokee County was also his place of residence. Broyles signed his parole No. 358 on July 31, 1865, at Marshall, Texas, as a 2nd lieutenant. Probably was so promoted after the February 29, 1864, muster roll on which he is shown as a private.

Buchaltz, James M., enlisted in Smith County, Texas, on June 10, 1863. Was sick in the hospital in Hempstead, Texas, at the time of the February 29, 1864, muster roll.

Calhoun, A.J., enlisted in Cherokee County, Texas, on February 10, 1863. Was with the horses near Fayetteville, Texas, on February 29, 1864.

Calhoun, W.D., enlisted in Cherokee County, Texas, on January 4, 1864. Discharged March 17, 1864, on a surgeon's certificate.

Churchwell, George W., enlisted in Cherokee County, Texas, on August 25, 1863. Wounded in Louisiana and was in the Keatchie, Louisiana, hospital on April 17, 1864.[46]

Cleveland, sick on a surgeon's certificate. No other information shown.

Coey, David W., enlisted in Smith County, Texas, on July 25, 1863. On February 12, 1864, he was detailed for duty in the Quartermaster Department in Houston.

Coker, Joseph, enlisted in Wood County, Texas, on April 9, 1863. He joined this company at Camp Lubbock by transfer in exchange for Private Rice Wells of Likens' regiment. Was an ambulance driver in December, 1863.

Coopland, James M., enlisted in Cherokee County, Texas, on February 10, 1863. Was sick at home at the time of the February 29, 1864, muster roll.

Curry, William D., enlisted in Colorado, Texas, on September 11, 1863. Was on detached service with the horses near Fayetteville, Texas, at the time of the February 29, 1864, muster roll.

Davis, James M., enlisted April 4, 1863. No other information found.

Gilbreath, A.J., enlisted April 3, 1863. Age 27, 5'8'' tall, hazel eyes, dark complexion, black hair. Civilian occupation, farmer. Received a disability discharge at Camp Groce, Texas, on July 31, 1863. Elbow had been injured in a fall from a horse.

Griffin, Andrew, enlisted in Cherokee County, Texas, on February 10, 1863. Was sick at a private house near Columbus, Texas, on December 20, 1863. Sick at the horse camp near Fayetteville, Texas, at the time of the February 29, 1864, muster roll.

Harris, William, enlisted in Smith County, Texas, on September 28, 1863. Was with the horses near Fayetteville at the time of the February 29, 1864, muster roll.

Hawkins, William F., enlisted in Smith County, Texas, on July 1, 1863. With the horses near Fayetteville, Texas, on February 29, 1864.

High, James J., enlisted in Van Zandt County, Texas, on April 20, 1863.

Hill, Patterson J., enlisted in Van Zandt County, Texas, on April 11, 1863. With the horses near Fayetteville, Texas, on February 29, 1864.

Hill, James, enlisted in Gilmer, Texas, on April 1, 1863. Age 20, 5'8" tall, blue eyes, fair complexion, dark hair. Civilian occupation, farmer. Born Caddo Parish, Louisiana. Discharged at Camp Groce, Texas, of chronic hepatitis.

Jones, Hamilton J., enlisted June 10, 1863. No other information shown.

Jones, R.J., Absent, sick. No other information shown.

Kerr, D.L., (could be Keir), transferred to McMahan's artillery battery on February 3, 1864, at Camp Sidney Johnston.

Kenard, Taylor E., enlisted in Cherokee County, Texas, on February 10, 1863.

Lakey, J.M., enlisted in Matagorda County, Texas, on December 10, 1863.

Langford, James M., enlisted in Cherokee County, Texas, on February 10, 1863. Was with the horses near Fayetteville, Texas, on February 29, 1864.

Lee, James C., enlisted in Smith County, Texas, on April 17, 1863.

Lewis, George P., enlisted in Van Zandt County, Texas, on April 1, 1863. Detailed in the Quartermaster Department in Houston as of February 12, 1864.

Little, Riley, enlisted in Cherokee County, Texas, on August 10, 1863. Absent, sick in the hospital.

Little, W.A., enlisted in Cherokee County, Texas, on February 10, 1863.

Martin, John, enlisted in Brazoria County, Texas, on December 15, 1863. Discharged March 31, 1865, from the general hospital in Houston of chronic bronchitis. Apparently did not go with the regiment to Louisiana as he had been detailed to the government shoe shop at Hempstead on January 5, 1864.

Mayfield, Asa, enlisted in Upshur County, Texas, on August 6, 1863. Was with the horses near Fayetteville, Texas, on February 29, 1864.

Mayfield, Hey, joined by enlistment at Camp Wharton, Texas, on December 20, 1863.

Mayfield, J.F., enlisted in Upshur County, Texas, on February 10, 1863.

McPherson, John F., enlisted in Van Zandt County, Texas, on April 18, 1863.

Moore, D.D., enlisted in Cherokee County, Texas, on September 1, 1863. Was with the horses near Fayetteville, Texas, on February 29, 1864.

Morgan, James E., enlisted in Cherokee County, Texas, on February 10, 1863.

Morrison, James P., enlisted in Cherokee County, Texas, on February 10, 1863.

Norman, W.Z., enlisted in Cherokee County, Texas, on January 4, 1863.

O'Kelley, James S., enlisted in Wood County, Texas, on March 31, 1863. Was admitted to the CSA Shreveport Hospital in Shreveport, La., on May 30, 1864. Apparently wounded. Died of gangrene on May 31, 1864.

Patterson, Major, enlisted in Prairieville, Texas, on December 15, 1862. Was 5'8'' tall, blue eyes, dark complexion, light hair. Born Pickens County, Alabama. Discharged July 31, 1863, at Camp Groce, Texas, of chronic hepatitis.

Pearce, William K., no information except that he was AWOL in November, 1863.

Pierce, E.D., enlisted in Cherokee County, Texas, on January 4, 1863. Was with the horses near Fayetteville, Texas, on February 29, 1864.

Pierce, George L., enlisted in Cherokee County, Texas, on January 4, 1863. Died at Camp Sidney Johnston on February 4, 1864.

Pearce, Richard, enlisted in Cherokee County, Texas, on February 10, 1863.

Pierson, M.O., no information except that he was sent after a horse in November, 1863.

Prickett, Joseph B., enlisted in Van Zandt County, Texas, on April 17, 1863.

Rather, C.C., enlisted in Cherokee County, Texas, on February 10, 1863. Absent, sick. Left on the Caney River.

Reaves, Jerry E., enlisted in Cherokee County, Texas, on February 10, 1863.

Renfrow, W.W., enlisted in Cherokee County, Texas, on September 1 1863.

Richards, James W., enlisted in Wood County, Texas, on April 3, 1863.

Robins, J.R., no information, except that he was on daily duty as a pioneer in December, 1863.

Robinson, W.P., enlisted in Cherokee County, Texas, on February 10, 1863.

Salome, Thomas, no information, except that he was on daily duty as a pioneer in December, 1863.

Self, E.B., furloughed at Camp Lubbock for 12 days to go home for a horse. Was dropped as a deserter on the return for December, 1863.

Self, E.H., enlisted in Cherokee County, Texas, on February 10, 1863. Dropped from the roll as a deserter February 1, 1864.

Self, John C., enlisted in Cherokee County, Texas, on January 11, 1864.

Sinclair, William, no information, except that he was absent without leave in November, 1863.

Sloan, Thomas, enlisted in Colorado County, Texas, on September 8, 1863. Signed parole No. 488 at Millican, Texas, on July 10, 1865.

Stockton, George R.L., enlisted in Cherokee County, Texas, on January 4, 1863.

Talley, Riley, no information except that he was sick on a surgeon's certificate, but was considered a deserter on December 5, 1863.

Taylor, George H.F., enlisted in Cherokee County, Texas, on January 4, 1863.

Taylor, Perry F., enlisted in Wood County, Texas, on April 3, 1863. Absent, sick in the hospital in Houston on February 29, 1864.

Walker, R.R., enlisted in Cherokee County, Texas, on February 10, 1863.

Wallace, Samuel M., enlisted in Smith County, Texas, on August 1, 1863.

Walters, Moses, enlisted in Smith County, Texas, on April 17, 1863. Was with the horses near Fayetteville, Texas, on February 29, 1864.

Wells, Rice, was transferred to Liken's regiment in exchange for Private Coker.

Whitman, Mertice J., enlisted in Smith County, Texas, on April 13, 1863. Whitman was born in Harris County, Ga., on May 1, 1845, and came to Smith County, Texas, in 1858 with his parents, being raised on a farm near Starrville in Smith County. He served in two other organizations before joining Terrell's regiment in June, 1863. After the war, he practiced law at Rusk, Texas, until November, 1876, when he was elected county attorney of Cherokee County. In 1882 he was elected county judge and served until 1890.[47]

Williams, Darlin M., enlisted in Cherokee County, Texas, on February 10, 1863.

Williams, Francis N., enlisted in Cherokee County, on August 20, 1863. Age 22, 5'6'' tall, blue eyes, fair complexion, dark hair. Civilian occupation, farmer. He applied for a furlough in order to go home and arrange for someone to look after his children and his property. His wife apparently died about December 24, 1863.

Williams, Thadeus S., enlisted in Cherokee County, Texas, on February 10, 1863.

Wood, James M., enlisted in Cherokee County, Texas, on February 10, 1863.

Yergan, John, enlisted in Van Zandt County, Texas, on August 1, 1863.

TERRELL'S TEXAS CAVALRY REGIMENT

GRAY'S COMPANY

CAPTAINS

Gray, J.E.

LIEUTENANTS

Gilmore, Stephen S.

PRIVATES

Arnette S.
Bates, James
Bowers, S.
Bresbon, W.
Brislow, W.
Butes, James
Campbell, A.W.
Dittmer, C.
Farmer, R.
Ferguson, J.H.
Fischer, J.F.
Givens, W.H.
Grimes, W.
Henotte, C.
Hodde, F.
Hughes, B.S.
Kumaka, H.
Manke, T.
McGregor, D.
McNeese, D.F.
McNeese, T.W.
Morphis, J.M.
Morris, E.H.
Ormand, J.B.
Pearson, S.J.
Ringner, H.
Seibill, A.

MULLINS' COMPANY

CAPTAINS

Mullins, W.H.

PRIVATES

Amos, William
Bone, J.T.
Brewers, C.C.
Brown, C.R.
Brown, T.
Brown, William
Burke, L.P.
Burris, A.J.
Davidson, T.S.
Dowdy, W.D.
Dukes, William
Frazier, J.
Hammand, J.W.
Hester, Z.
Holcomb, H.H.
Holmes, A.J.
Jenkins, M.A.
Jones, A.
Luce, William
Maze, J.
McAnully, E.P.
Moore, O.B.
Nelson, Henry F.
Pinson, A.
Reasons, Jr. P.
Red, N.H.
Shoffit, W.A.
Smith, William
Smith, J.
Squire, J.
Walker, J.
Walker, R.
Wheeler, A.W.
Williamson, S.
Wirthey, J.T.

FOOTNOTES

[1]The names for this roster were copied from the names shown on *Micro Copy No. 323,* The National Archives, National Archives and Records Service, General Services Administration, Washington, D.C., 1960, Rolls No. 177, 178, and 179. The rolls studied were at the Dallas Public Library, Dallas, Texas, Genealogy Section. The author has in his possession a microfilm of the muster rolls from which the above microfilm were made.

[2]*Alexander Watkins Terrell Papers,* University Archives, Barker Texas History Center.

[3]The Galveston *Tri-Weekly News,* April 25, 1864.

[4]L.E. Daniel, *Personnel of the Texas State Government,* San Antonio, Maverick Printing House, 1892, 205-206. Microfilm 323, *8th Texas Cavalry,* Roll 51.

[5]*The Lone Star State,* Lewis Publishing Co., 1893. (From a 1966 reprint arranged and indexed by Stephenie H. Talley-Frost). p. 77.

[6]Card index of graves of Civil War soldiers in Navarro County, Corsicana Public Library, Corsicana, Texas.

[7]The Galveston *Tri-Weekly News,* April 25, 1864.

[8]The Galveston *Tri-Weekly News,* April 25, 1864.

[9]Sidney Smith Johnson, *Texans Who Wore the Gray,* Tyler, Texas, 1907, p. 267.

[10]Margaret Horn, *Introduction to the History of Hunt County, Greenville, Texas,* (No date), unpublished manuscript on file in the W. Walworth Harrison Public Library, Greenville, Texas, 17-18.

[11]Bee, "Battle of Pleasant Hill — An Error Corrected," *Southern Historical Society Papers,* 186.

[12]Microfilm 323, *Terrell's Texas Cavalry,* Roll 178. Personal interview with grandson, Mr. Robert Hood of Breckenridge, Texas, and granddaughters, Mrs. Dorothy Dennison, Amarillo, Texas, Mrs. Addie Simmons, Breckenridge, Texas, and Mrs. Theda Hood Spencer, Breckenridge, Texas.

[13]The Houston *Daily-Telegraph,* June 8, 1864.

[14]The Galveston *Tri-Weekly News,* April 25, 1864.

[15]*George Washington Diamond's Account of The Great Hanging at Gainesville, 1862,* ed. Sam Acheson and Julie Ann Hudson O'Connell, 1.

[16]Houston *Daily-Telegraph,* June 8, 1864.

[17]The Galveston *Tri-Weekly News,* April 25, 1864.

[18]The Galveston *Tri-Weekly News,* April 25, 1864.

[19]The Galveston *Tri-Weekly News,* April 25, 1864.

[20]The Galveston *Tri-Weekly News,* April 25, 1864.

[21]Houston *Daily Telegraph,* June 8, 1864.

[22]Houston *Daily Telegraph,* June 8, 1864.

[23]Card index of graves of Civil War soldiers in Navarro County, Corsicana Public Library, Corsicana, Texas.

[24]The Galveston *Tri-Weekly News,* April 25, 1864.

[25]Houston *Daily-Telegraph,* June 8, 1864.
[26]Houston *Daily-Telegraph,* June 8, 1864.
[27]The Galveston *Tri-Weekly News,* April 25, 1864.
[28]The Galveston *Tri-Weekly News,* April 25, 1864.
[29]Houston *Daily Telegraph,* June 8, 1864.
[30]The Galveston *Tri-Weekly News,* April 25, 1864.
[31]The Galveston *Tri-Weekly News,* April 25, 1864.
[32]The Galveston *Tri-Weekly News,* April 25, 1864.
[33]Johnson, *Texans Who Wore the Gray,* 75.
[34]Johnson, *Texans Who Wore the Gray,* 35.
[35]Johnson, *Texans Who Wore the Gray* 22.
[36]The Galveston *Tri-Weekly News,* April 25, 1864.
[37]The Galveston *Tri-Weekly News,* April 25, 1864.
[38]The Galveston *Tri-Weekly News,* April 25, 1864.
[39]The Houston *Daily-Telegraph,* April 25, 1864.
[40]The Galveston *Tri-Weekly News,* April 25, 1864.
[41]The Houston *Daily-Telegraph,* April 25, 1864.
[42]The Houston *Daily-Telegraph,* June 8, 1864.
[43]The Houston *Daily-Telegraph,* April 25, 1864.
[44]The Galveston *Tri-Weekly News,* April 25, 1864.
[45]The Houston *Daily-Telegraph,* June 8, 1864.
[46]The Houston *Daily-Telegraph,* April 25, 1864.
[47]Johnson, *Texans Who Wore the Gray,* 195-196.

Appendix IV

Likens' Texas Cavalry Regiment
(35th Texas Cavalry)

Likens' 35th Texas Cavalry regiment was organized October 23, 1863, by consolidation of Likens' and Burns' battalions. In a letter from Likens to Captain E.P. Turner, A.A. General, Houston, dated October 7, 1863, Likens stated that the regiment was organized with ten companies per Special Order No. 168. Three companies were then in Robertson County, five companies were near Tyler in Smith County, one company was in Cherokee County, and one company was in Houston County.[1] Later, in the reorganization of Terrell's regiment and Likens' regiment, Captains Gray's, Mullins', and Hurley's companies of Terrell's regiment were transferred to Likens' regiment by Special Order No. 321 dated November 25, 1863.[2]

Likens' battalion, which eventually formed the nucleus of the regiment when it was formed, was apparently organized very early. In a letter addressed to Likens dated November 9, 1861, Likens, then a major commanding a battalion at Sabine Pass, Texas, was instructed to be extremely vigilant in order to foil any attempt on the part of the Federals to surprise him. Likens was instructed to make a daily report to the commanding general of all information obtained by his scouts, and to send the dispatches to the point where they would be met by scouts from Bolivar Point. The envelope was to be endorsed by Likens "To be forwarded instantly on its receipt, day or night," and Likens was to be particular in his instructions to the bearer of the dispatches. Likens was to have scouting parties out twenty-four hours a day, and were to be strung out along the beach from Sabine to a point where they would meet scouts from the Galveston area. Likens was also to inform the commanding general of how many companies he had mustered into the service, and the progress of those companies which were in the process of formation.[3]

On December 16, 1861, Mr. A.M. Gentry, president of the Texas and New Orleans Railroad Company, wrote General P.O. Hebert, commanding general of the Department of Texas, that Major Likens, commanding at Sabine Pass, had cooperated with him in moving the rolling stock of the railroad up to a point of temporary safety.[4]

The 35th Cavalry served on the Texas coast during late 1863 and early 1864 along with most of the other regiments which eventually would fight in the Red River Campaign in Louisiana. In early December, 1863, the regiment was attached to Luckett's brigade and ordered to move as rapidly as possible to Velasco, leaving their horses behind. Federals were moving to this point by the beach from Saluria. If the Federals had already captured Velasco, Likens was to move on to Galveston.[5] On December 6, 1863, the regiment was ordered to march across the Colorado River at Elliott's Ferry and join Colonel Duff, who was either at Texana, or on the march near Victoria.[6] On December 7, 1863, it was reported that Colonel Likens had been arrested by Ashbel Smith of Luckett's brigade for being intoxicated for the past twenty-four hours and unable to perform his duties.[7] By January 9, 1864, Likens' regiment was under the command of Colonel A. Buchel at a camp near P. McNeil's farm near the mouth of the Caney River. Five companies of the regiment were ordered along with Colonel Likens to move to supporting distance of Colonel Buchel, who was observing Federal gunboats and transports.[8]

By January 22, 1864, Likens' regiment had been remounted and was with General Bee at Ewing's Plantation, near the mouth of the Caney. A large part of the regiment under Colonel Likens was sent down the beach with orders to keep under observation a large force of Federals who had landed and were marching southward from the mouth of the Caney. Likens had orders to return before daylight on January 22, which he did without ever overtaking the Federals.[9] On January 23, Likens took thirty men on another scouting expedition south of the mouth of the Caney. Upon his return on January 24, he reported that the Federals were still moving southward, their rear covered by three gun boats. The force of Federals was estimated by Likens to be 2,500 strong, with three pieces of artillery, and one hundred cavalry troops. Likens thought that their objective must be to get a meat supply for the boats.[10]

By January 22, word of the large force of Union troops which had arrived at the mouth of the Mississippi River had reached Texas. Initial thoughts were that the Federals were destined for Galveston, so all troops were alerted to be ready to move toward that city for its defense.[11] Likens' regiment was one of those alerted to be ready to move to Galveston.[12] On January 31, 1864, Likens' regiment was reported to be in camp near San Bernard, along with the 2nd Texas Cavalry and Colonel P.C. Woods' 36th Texas Cavalry.[13]

The 35th was placed in a division of cavalry troop along with Terrell's, Buchel's, Debray's, Woods', and Gould's for movement to Louisiana. Commander of the division was H. P. Bee, who marched with the six regiments on March 11, 1864, toward Louisiana, arriving at Mansfield with only three of them, Terrell's, Buchel's, and Debray's. Likens' regiment, along with those of Gould and Woods being delayed by incidents of the long march, did not arrive in time for the battles of Mansfield and Pleasant Hill.[14]

On March 31 Debray's regiment crossed the Sabine River, and Buchel's regiment crossed the following day on April 1, but Likens' regiment was still at least two days away.[15] Apparently Likens' regiment was more than two days away as an urgent message was sent to him on April 2 from the assistant adjutant general at Sabinetown directing him to deposit all baggage that could be spared in order to make a quick and rapid march, and to hasten on as rapidly as possible. In the message it was pointed out that the troops would soon have an opportunity to distinguish themselves, and that Likens' regiment was wanted in the fight. Likens' was ordered to proceed as rapidly as his horses would travel, bringing along all his wagons loaded with provisions and forage.[16]

Had the regiment arrived in time, it probably would have gone with Woods' and Gould's regiments with General Tom Green to Blair's Landing, where General Green was killed on April 12. Therefore, it appears that the regiment did not arrive in Louisiana until sometime after the Blair's Landing battle. On April 21, General Bee was ordered to take his division of cavalry to the Red River to prevent the passage of the Federal transports with supplies destined for the Union army.[17] Bee does not mention anything about being short one regiment, giving the impression that he had his entire six regiments with him at this time, namely, Terrell's, Debray's, Buchel's, Woods', Gould's, and Likens'.

Colonel Terrell appears to have been in command of a brigade of three regiments at Monett's Ferry, his own 34th Texas Cavalry, Yager's 1st Texas Cavalry, and Likens' 35th Texas Cavalry. General Richard Taylor mentions in his report of April 24, 1864, that Bee was at Monett's Ferry with the brigades of Major's, Bagby's, Debray's, and Terrell's cavalry, implying that Terrell had three regiments wth him, rather than just two. Taylor further points out that one of Bee's errors in the Monett's Ferry affair was, first of all in sending Terrell's entire brigade to Beasley's Station to look after the subsistence train, which Taylor says that he had already taken care of adequately.[18]

On April 26, 1864, Likens' regiment attacked four Federal gunboats and two transports at Montgomery. One of the gunboats had been unable to pass the bar below Montgomery and the other boats had remained behind to assist it. Likens' troops opened fire about 9:00 a.m. killing many of the men working to get the large gunboat over the bar. After some delay the gunboats opened a heavy fire on the Confederates and commenced moving down the river again. The large gunboat struck the bar, the Federals blew it up, and continued down the river.[19]

Several cavalry regiments were dismounted in early 1865 and Likens' regiment was one of them.[20] The regiment started for Texas on about February 19, 1865.[21] Bagby's (Herbert's) old 7th Texas Cavalry regiment replaced Likens' regiment in Terrell's brigade.[22] Likens' regiment was to join the newly organized infantry division of Major General Sam B. Maxey. It would be assigned to the 1st Brigade along with another recently dismounted cavalry regiment, Hardeman's (Colonel Peter Hardeman), and two infantry regiments, the 15th and the 17th.[23] Several changes were apparently necessary in the organization of Maxey's division of infantry after the initial assignment of Likens' regiment to it. Special Order No. 118 of the District of Texas, New Mexico, and Arizona dated April 28, 1865, gave a new and final arrangement of Maxey's division. Likens' regiment was not mentioned as being a part of that arrangement.[24] Apparently the regiment was needed more at Galveston. An abstract from the field return of the defenses of Galveston, with Colonel Ashbel Smith commanding, dated May 10, 1865, showed the 35th Texas Dismounted Cavalry as a part of that command. Present for duty were fifteen officers and 165 men, total present, 232 men.[25] There had been a serious mutiny at Galveston, a recurrence of which was expected, and Likens' regiment was probably needed to help restore order.[26]

By May 21, 1865, Likens' regiment along with the others on Galveston Island were being told that they would all receive honorable discharges if they behaved properly. Orders were being issued to move the troops west by regiments to Harrisburg, where they could get supplies and board trains to go to their homes. Their discharges were already made out and waiting to be given to them. Instructions were given to spike all the guns that could not be taken with them, and to take the telescope from the observatory. The regiment probably started moving on about May 25 toward Harrisburg. They were probably all together there for one or two days, and if any waited long enough, were probably given the promised discharges. The regiment probably broke up on about May 27, 1865, at Harrisburg, Texas.[27]

It appears that Likens' regiment was brigaded with Colonel Terrell's regiment and Yager's regiment from its arrival in Louisiana on about April 15, 1864, to its departure on February 19, 1865. During this time it would have been engaged in all the skirmishes and battles along with Terrell's regiment. The regiment apparently fought in the battles of Mansura and Yellow Bayou, as well as the skirmish on September 21 at Morgan's Ferry. On the retreat from this skirmish Likens' regiment destroyed a Federal pontoon bridge on the Bayou Rouge.[28]

Having missed the bloodiest battles of the Red River Campaign, Likens' regiment had rather light casualties. The regiment had only two wounded at the Yellow Bayou battle, and suffered only five wounded, with no killed or missing in the entire Red River Campaign. By comparison, Terrell's regiment had five killed, twenty-six wounded, and ten missing in the campaign; and Yager's regiment had nine killed, fifty-nine wounded, and four missing.[29]

Unfortunately there are no muster rolls on file for Likens' regiment. Apparently they were lost in the confusing times at the end of the war, in either Shreveport or Houston, where the Trans-Mississippi Department made its headquarters. Fortunately, however, the following record of the various officers of the regimental field and staff and company officers were preserved.[30]

LIKENS' TEXAS CAVALRY REGIMENT
(35th Texas Cavalry)

FIELD AND STAFF

Likens, James B., Colonel. His parole was signed on July 12, 1865, at Sabine, Texas. Signed the parole as "Colonel, 35th Regt., Texas Dismounted Cavalry."

Burns, J.R., Lt. Colonel. Resigned 1864.

Wortham, W.A., Major and Lt. Colonel.

Cameron, B.F., A.Q.M.

Seales, J.R., S (Probably surgeon).

Haynes, D.R., A.S. (Probably assistant surgeon). Resigned, no date shown.

Beynon, William, Adjutant.

Barkley, W.A. (or Baxley), Adjutant.

COMPANY A

Black, J.N. Captain, resigned on account of ill health while at Evergreen La., on September 26, 1864.
McLendon, J.S., 1st Lieutenant.
Evans, William, 2nd Lieutenant. Resigned 1864.
Tomlinson, A.A., 2nd Lieutenant.
Company A was organized on September 12, 1863.

COMPANY B

Rawson, Henry B., Captain.
Coleman, Thompson H., 1st Lieutenant.
Clough, George W., 2nd Lieutenant, resigned.
Hudson, A.B., 2nd Lieutenant, resigned 1864.
Company B was organized on September 19, 1863.

COMPANY C

Dawson, William W., Captain.
McKnight, Joseph M., 1st Lieutenant.
Luther, Elbert S., 2nd Lieutenant, resigned 1864.
Chenoweth, John W., 2nd Lieutenant, resigned 1864.
Company C was organized on August 29, 1863.

COMPANY D

Gray, James E., Captain, resigned.
Gilmore, S.H., 1st Lieutenant, resigned.
Hosea, G.W., 2nd Lieutenant.
Blankenship, H.H., 2nd Lieutenant, resigned 1864.
This company was formerly Gray's Company of Terrell's regiment and was organized in July, 1863. A partial roster of this company is shown immediately following Terrell's regiment.

COMPANY E

Hickey, Granville, Captain.
Boyd, H.A., 1st Lieutenant.
Fields, D.M., 2nd Lieutenant.
Rose, G.W., 2nd Lieutenant.
Company E was organized on June 1, 1863.

COMPANY F

Wiggins, J. T., Captain.
Murry, John, 1st Lieutenant.
Roberts, William, 2nd Lieutenant.
Park, James T., 2nd Lieutenant.
Company F was organized on October 21, 1863.

COMPANY G

Warren, G. E., Captain.
Ford, R. W., 1st Lieutenant.
McGraw, E. W., 2nd Lieutenant.
Gillespie, A. T., 2nd Lieutenant, resigned.
Johnson, W. R., 2nd Lieutenant.
Company G was organized on September 19, 1863.

COMPANY H

Bates, G. W. Captain. Signed his parole at Marshall, Texas, on July 17, 1865. His residence was shown as Smith County, Texas.
Wiggins, Thomas, 1st Lieutenant.
Jones, W., 2nd Lieutenant.
Dui, Thomas J., resigned January 7, 1865.
Company H was organized October 7, 1863.

COMPANY I

Mullins, W. H., Captain.
Moore, J. W., 1st Lieutenant.
Lunsford, J. H., 2nd Lieutenant.
Frazier, John J., 2nd Lieutenant.
This company was formerly Mullins' company of Terrell's regiment. See the partial roster following Terrell's regiment.

COMPANY K

Boren, A. B., Captain. Signed his parole at Marshall, Texas, on July 11, 1865. Residence, Upshur County, Texas.
Trowell, John H., 1st Lieutenant. Formerly in Terrell's Companies K and D.

Whitworth, W. L., 2nd Lieutenant.
Rogers, A. C., 2nd Lieutenant.
This company was organized from Hurley's company of Terrell's regiment and the men of T. N. George in November, 1863. Hurley's company was originally organized on April 1, 1863. Hurley lost out in the election of officers.

FOOTNOTES

[1] Microfilm 323, *Likens' Texas Cavalry,* Roll 171.

[2] Microfilm 323, *Likens' Texas Cavalry,* Rolls 171, 177.

[3] *Official Records,* IV, 134-135.

[4] *Official Records,* IV, 158.

[5] *Official Records,* XXVI, Part 2, 472.

[6] *Official Records,* XXVI, Part 3, 487-488.

[7] Microfilm 323, *Likens' Texas Cavalry,* Roll 171.

[8] *Official Records,* XXXIV, 80-81.

[9] *Official Records,* XXXIV, Part 2, 906.

[10] *Official Records,* XXXIV, Part 2, 912-913.

[11] *Official Records,* XXXIV, Part 2, 907.

[12] *Official Records,* XXXIV, Part 2, 913.

[13] *Official Records,* XXXIV, Part 2, 933.

[14] *Official Records,* XXXIV, 606.

[15] *Official Records,* XXXIV, Part 3, 722.

[16] *Official Records,* XXXIV, Part 3, 725.

[17] *Official Records,* XXXIV, 610.

[18] *Official Records,* XXXIV, 580.

[19] *Official Records,* XXXIV, 583.

[20] *Official Records,* XLVIII, 1392.

[21] Bitton, Ed., *The Reminiscences and Civil War Letters of Levi Lamoni Wight,* 176.

[22] *Official Records,* XLVIII, 1390; Bitton, ed., *The Reminiscences and Civil War Letters of Levi Lamoni Wight,* 178.

[23] Alwyn Barr, *Polignac's Texas Brigade,* Texas Gulf Coast Historical Association, Vol. VIII, No. 1, November, 1964, 54.

[24] *Official Records,* XLVIII, Part 2, 1291.

[25] *Official Records,* XLVIII, Part 2, 1297.

[26] *Official Records,* XLVIII, Part 2, 1300.

[27] *Official Records,* XLVIII, Part 2, 1316-1317.

[28] *Official Records,* XLI, 810-812.

[29] Alwyn Barr, "Texas Losses in the Red River Campaign, 1864," *Texas Civil War Centennial Program for the Centennial Commemoration of the Red River Campaign,* Odessa, Texas, West Texas Office Supply, A Publication of the Texas Civil War Centennial Commission and Texas Historical Survey Committee, 1962, 35.

[30] Microfilm 323, *Likens' Texas Cavalry,* Roll 171.

Appendix V

Yager's Texas Cavalry Regiment (1st Texas Cavalry)

This regiment of Texas Cavalry was formed about May, 1863, by the consolidation of the 3rd (Yager's) Battalion Texas Cavalry and 8th (Taylor's) Battalion Texas Cavalry, and Captain Ware's company of Texas cavalry.[1] The regiment served along the Texas coast until March, 1864, when it was started on the march to Louisiana along with Terrell's regiment and the other cavalry units of Texas which participated in the Red River Campaign. At one time Terrell served in this regiment as a major.[2] General Richard Taylor was very impressed with the drill and discipline of this regiment, being careful to call them "cavalry" to distinguish them from "horse," which he called mounted infantry.[3]

From the time the 1st Texas Cavalry joined General Taylor at Mansfield on April 5, 1864, until the return to Texas with Terrell's brigade, the regiment served gallantly, suffering many casualties. In the battles of Mansfield and Pleasant Hill alone the regiment had nine killed, fifty-one wounded, and four missing.[4] No doubt many of these casualties occurred during the battle of Pleasant Hill and the furious fighting after the Yankee ambush of the 1st Texas and Debray's 26th Texas. The regiment was heavily engaged in the third phase of the battle of Mansfield after they made their way through the dense woods on the right of the Confederate line along with Debray's cavalry. General Bee called this battle the battle of the Peach Orchard, and said that it was a distinct battle from the battle of Mansfield. The 1st Texas Cavalry and Debray's cavalry were not engaged until after the first two phases, and after the capture of the Federal wagon train and artillery by the infantry and other cavalry who had fought as infantry.[5] Private Levi Wight of the 1st Texas stated that the Peach Orchard was the bloodiest of the three phases.[6]

Colonel Debray's 26th Texas was the lead regiment in the charge at Pleasant Hill, with the 1st Texas following close behind. The mission was to move forward, form a line of battle, and charge through the village of Pleasant Hill on the Federals. Colonel Buchel was able to draw the 1st Texas back in time to avoid the ambush, passed to the left of Debray's regiment, dismounted, and attacked the Federals, driving them back to their lines. It was during this action that Colonel Buchel was mortally wounded.[7]

Lt. Colonel Yager was promoted to colonel after the battle of Pleasant Hill and would remain in command until the end. Colonel Yager now led the regiment, along with the rest of the division of cavalry under General Bee, and a battery of artillery to Monett's Ferry to attempt to cut off the retreat of the Union army. Another Confederate division of cavalry under General Major was already in position at the ferry crossing, and General Taylor hoped to encircle the Federal army and capture them all. Both Terrell's and Yager's regiments were sent on to Beasley's Station, where a supply camp had been established and were, therefore, not engaged in the action at Monett's Ferry.[8]

On May 17, Colonel Yager led the 1st and probably the 35th (Likens' regiment) against the Federal wagon train near Moreauville, killing and driving off the guard and destroying much property.[9]

The 1st Texas Cavalry continued in action almost daily, along with the rest of the Texas troops in Louisiana, until the Federal army finally completed their escape across the Atchafalaya River on May 20, 1864, and the end of the Red River Campaign.

Yager's 1st Texas Cavalry was placed in a brigade of cavalry under Colonel Terrell in early September, 1864.[10] The brigade was on the line of the Atchafalaya River somewhere between Simmesport and Morgan's Ferry. The regiment participated in the fight at Morgan's Ferry on September 17 and September 20. The regiment remained with Colonel Terrell during the patrol of the Atchafalaya River and their duty on the Bayou Boeuf. The winter of 1864-1865 was passed at Alexandria, with patrols being sent out to combat jayhawkers.

On about April 1, 1865, all the men of the brigade who were without horses were given furloughs to return home to Texas and remount themselves. Private Levi Wight of Company K in the 1st Texas Cavalry said that:

> "Many soldiers in our command were without horses. All that were without horses were to get furloughs for sixty days to remount themselves. Most all had horses at home. I and the rest that were equipped as the law directed were out this time, but we fell to trading. I was successful in exchanging my horse with a young man for his furlough."[11]

Wight and the other members of the brigade who left the regiment at that time to remount themselves would never again see the brigade as the war would end and the regiment would disband at Wild

Cat Bluff on the Trinity River in Texas along with the regiments of Terrell, commanded by Lt. Colonel Robertson, and Herbert's 7th Texas Cavalry Regiment.

YAGER'S TEXAS CAVALRY REGIMENT[12]
(1st Texas Cavalry)

FIELD AND STAFF

Buchel, Augustus, Colonel. Mortally wounded at Pleasant Hill, Louisiana, April 9, 1864, and died two or three days later in the camp of General H. P. Bee.
Yager, William O., Lt. Colonel, promoted to colonel effective April 13, 1864, after the death of Buchel. Signed his parole at San Antonio, Texas, on September 2, 1865. Residence at the time of his parole was shown as Seguin, Texas.
White, James A., probably subsistence officer. Was dropped February, 1864.
White, John L., probably assistant subsistence officer. Was dropped February, 1864.
Jordan, Charles H., Assistant Subsistence Officer.
These are the only names readable on the microfilm.

COMPANY A

Brewin, William H., Captain, resigned December, 1863.
Bass, Charles C., 1st Lieutenant, Captain.
Cunningham, James J., 1st Lieutenant, resigned February, 1864.
Williams, John E., 2nd Lieutenant, deserted February, 1864.
Emmons, James D., 2nd Lieutenant, deserted February, 1864.
Van Allen, George W., 1st Lieutenant.
Ferguson, Joseph J., 2nd Lieutenant.

COMPANY B

Beaumont, Edward, Captain, promoted to Major.
Cunningham, Harvey S., 1st Lieutenant, Captain.
Blair, Thomas A., cashiered by general court-martial.
Patton, Isaac A., 1st Lieutenant.

COMPANY C

Donnelly, Batholomeu, resigned February 27, 1865.
Vincent, James C., 1st Lieutenant.
Duckworth, William T., 2nd Lieutenant.
Woodlief, M., 2nd Lieutenant, resigned 1864.

COMPANY D

Boren, James C., Captain. Severely wounded at the Peach Orchard in Louisiana.[13]
Flournoy, Rice C., 2nd Lieutenant.
Preston, Landon C., 2nd Lieutenant.
Bishop, Benj. F., 2nd Lieutenant.
This company was at one time attached to the 6th Texas Infantry.

COMPANY E

Stucken, Frank, Captain, resigned January, 1865.
Max, George, 1st Lieutenant.
Siemering, August, 2nd Lieutenant, resigned May, 1864.
Heffler, Hugo O., 2nd Lieutenant, resigned July 11, 1864.

COMPANY F

Ware, James A., Captain.
Maim, Walker L., 1st Lieutenant.
Shook, John R., 2nd Lieutenant, resigned May, 1864.
Murphy, Pat F., 2nd Lieutenant.

COMPANY G

Price, Leonidas, Captain.
Homsley, James M., 1st Lieutenant.
Bealty, Charles R., 2nd Lieutenant.
George, W.H., 2nd Lieutenant.

COMPANY H

Combs, Francis H., Captain.
Bushong, George E., 1st Lieutenant.

Williams, John S., 2nd Lieutenant.
Gracey, E. A., 2nd Lieutenant.

COMPANY I

No information is shown on the microfilm. The company was at Camp Sidney Johnston on the January and February, 1864, muster rolls.

COMPANY K

Bigham, James S., Captain.
Bell, James D., 1st Lieutenant.
Kuykendall, George R., 2nd Lieutenant.
Woodruff, D. W. 2nd Lieutenant.

FOOTNOTES

[1]Microfilm 323, *Yager's Texas Cavalry,* Roll 5.

[2]Harry McCorry Henderson, *Texas in the Confederacy,* San Antonio, The Naylor Company, 1955, 123-124.

[3]Taylor, *Destruction and Reconstruction,* 190.

[4]Barr, "Texas Losses in the Red River Campaign," 31.

[5]*Official Records,* XXXIV, 607.

[6]Bitton, Ed., *The Reminiscenses and Civil War Letters of Levi Lamoni Wight,* 90.

[7]*Official Records,* XXXIV, 608.

[8]Barr, ed., "Mechling's Journal," April 22; Johnson, *Red River Campaign,* 234.

[9]Taylor, *Destruction and Reconstruction,* 232; Johnson *Red River Campaign,* 274.

[10]*Official Records,* LI, 854.

[11]Bitton, ed., *The Reminiscences and Civil War Letters of Levi Lamoni Wight,* 93.

[12]Microfilm 323, *Yager's Texas Cavalry,* Roll 5.

[13]*Official Records,* XXXIV, 607.

Appendix VI

Herbert's Texas Cavalry
(7th Texas Cavalry)

The history of this regiment goes back to the organization of Sibley's Arizona Brigade, formed by Henry Hopkins Sibley in 1861 with orders by the Confederate government to invade the territory of Arizona, which is the present state of New Mexico. Sibley organized three regiments and called them the 4th, 5th, and 7th Texas Volunteer Cavalry regiments, respectively. The 7th regiment was commanded by Captain William Steele, who had been a captain in the U.S. Army. Headquarters for the brigade was in San Antonio and the regiments were in two camps on Salado Creek, which was about five miles from the city near the Austin highway crossing, and on the Leona River. The 7th had completed its organization by October 4, 1861, and spent its time before it departed for New Mexico in disciplining and drilling. Captain William P. Hardeman commanded the 4th regiment, and Thomas Green commanded the 5th regiment. After organization of the brigade, all would be made colonels of their respective regiments. Lt. colonel of the 7th was J. S. Sutton and major was A. P. Bagby.

Colonel Steele started the 7th from camp for New Mexico on November 20, 1861, and arrived at Fort Bliss on December 14. Five companies of the regiment marched north with Sibley during the first week in February, 1862, and by February 7 had reached to within seven miles of Fort Craig. After some preliminary movements and counter movements by the Federals, Sibley decided to force a battle at a mesa about seven miles above Fort Craig in what would be called the battle of Valverde. The 7th Texas fought valiantly in this battle.

The 7th Texas remained in New Mexico with Sibley until April 12, when they began marching out of Albuquerque, and eventually out of New Mexico. The brigade by this time was in extremely bad condition. They had casualties and sick with them, no food, their horses had long since been worn out, and they were now dismounted. Morale was low. They arrived back at Fort Bliss the first week in May, remained a few days, and started for San Antonio, in a disorganized march where it was every man for himself. On arrival at San Antonio the men were given furloughs, and upon return to camp at Hempstead, Texas, were reorganized and recruited back to full strength. Thomas J. Green was made commander of the new brigade, being promoted to brigadier general.

First action of the newly organized brigade was in the capture of Galveston on January 1, 1863. The 7th regiment furnished volunteers to help man two steamboats for that part of the attack on water while the rest of the regiment fought on land.

About the middle of February, 1863, orders were received which sent the brigade to Louisiana, where they made camp at Opelousas during the second week in March. They participated in a battle on Bayou Teche in southern Louisiana, forming the rear guard in the retreat from Camp Bisland. They participated in the capture of a Federal post at Berwick Bay and also in a battle of La Fourche on July 14, 1863. In September, 1863, Brigadier General Green was promoted to major general and given command of a division, which included Green's brigade, as it was now called, and Colonel Major's brigade. A.P. Bagby was now commander of the old brigade, and Lt. Colonel P.T. Herbert became commander of the 7th Texas Mounted Rifles.[1]

On November 3rd, 1863, the 7th Texas Cavalry, along with several other Texas regiments of cavalry and infantry, attacked the Federal rear guard at Bayou Bourbeau, with the 7th cavalry making a brilliant mounted cavalry charge, breaking the Federal line and forcing them to give way.[2]

Bagby's cavalry brigade left Louisiana and marched for Houston on December 14, 1863, arriving on Christmas Day. They remained in Texas until February 15, 1864, when they were again started on the march to Louisiana.[3]

The 7th Texas Cavalry was one of the first reinforcements received by General Taylor during the Red River Campaign, reporting to the general on the morning of March 31, 1864, at Natchitoches.[4] On April 2, the 7th and 5th Texas Cavalry regiments were on duty near Many when they encountered Federal cavalry. They fought until their ammunition was exhausted, then fell back toward Pleasant Hill. It was here that Debray's 26th Texas Cavalry had its baptism of fire as they had unexpectedly run into the rear of the Federals while on the way to report to General Taylor at Pleasant Hill.[5]

On April 7, the 7th Texas Cavalry, along with the 5th Texas Cavalry regiment, under Brigadier General Bagby, engaged in heavy skirmishing with Federal cavalry south of Pleasant Hill, skirmishing all the way back through that village near noon, and on back to Wilson's farm, about three miles from the village. Here they were reinforced by Lane's brigade, who had camped the previous night at Ten Mile Bayou. The 7th Texas and other Confederates drove the Federals back in unexpected force until the Federals received reinforcements, then they withdrew to Ten Mile Bayou, leaving skirmishers to slow down the advance of the enemy. Here the 7th and 5th regiments were reunited with the 4th regiment of Colonel Hardeman, which had arrived at Mansfield on April 5 and moved on to Ten Mile Bayou.

The Confederate skirmishers brought the Union troops on slowly, in a restraining fight, giving the other regiments of Bagby's and Lane's brigades time to form a battle line at Ten Mile Bayou. Here the 7th Texas, and the other two regiments of the brigade, formed the right wing of the Confederate force, and Lane's brigade formed the left wing, with the Valverde battery placed in the center of the two brigades. When the skirmishers arrived at the battle line, the Valverde battery opened up with grape and canister on the Federals, and a fierce cannonade lasting about an hour took place. The Federals crossed Ten Mile Bayou in front of the 7th Texas, but the 7th and the rest of the brigade poured such a heavy fire into them that they were forced to fall back, all the while absorbing the grape and canister of the Valverde battery. A second time they attempted to cross the little stream, and a second time they were repulsed.[6] By this time Terrell's regiment, along with Buchel's and Debray's regiments had joined the Confederate force, and night put an end to the fighting. The two forces bivouacked on opposite sides of the bayou while a slow rain began to fall on them.

At daylight the next morning the 7th Texas moved back to a point about three miles south of Mansfield, where General Taylor had decided to make a stand against the Union force. The regiment occupied the left side of the Confederate line under General Major, and fought as the rest of the cavalry did on that side of the line on foot.[7]

During the battle of Pleasant Hill, the 7th cavalry regiment occupied the left side of the Confederate line, next to the two regiments of General Bee, which were poised across the road, ready to charge through the village of Pleasant Hill, a charge which later proved to be premature and which cost many valuable lives, including the life of Colonel Buchel of the 1st Texas Cavalry regiment.[8] Their mission in the battle was to move forward, dismounted, against the Federal right wing, turning it inward, and hold the road to Blair's Landing.[9]

After the battle of Pleasant Hill, the 7th cavalry regiment was sent with the rest of the brigade commanded by Bagby toward Grand Bayou Landing with the objective of trying to intercept any messages which General Banks might try to send to the fleet on the river, and to do as much damage to the retreating fleet as possible. But to get to the river, Bagby's brigade had to cross the Bayou Pierre, 300 feet wide and too deep to ford. Without a pontoon bridge, the only means available was one small flat ferry boat, which use was slow and caused the brigade to fail in both of their objectives. This was on April 11, and the fleet passed Grand Bayou Landing at 10:00 a.m., several hours before Bagby's brigade was able to make it to that point. After missing the Federal fleet at Grand Bayou Landing, the 7th and the rest of the brigade pushed on toward Blair's Landing on the river, where they arrived on the night of April 12, too late to participate in the Battle of Blair's Landing, which had been fought earlier in the day, and which had cost the life of General Green, as well as many men of Woods' and Gould's regiment.

After the affair at Blair's Landing, the 7th regiment returned to Pleasant Hill, and then joined General Bee in front of Grand Ecore, where Banks had his army concentrated,[10] twenty-five thousand Federal troops hemmed in by about five thousand Confederates.[11]

Next action by the 7th Texas Cavalry was at Monett's Ferry on April 23, where General Bee was sent with about 2,000 cavalry troops to intercept Banks' army on the Cane River in their retreat southward toward Alexandria. Outflanked on both his right and left, and with a formidable force of 25,000 in front of him, Bee had no alternative but to give way, retreating to Beasley's Station where Terrell and Yager had been sent with their regiments.[12]

Leaving Beasley's Station on April 26, the 7th, along with the rest of the cavalry, pressed the rear of the retreating Federals. A brisk skirmish resulted at Bayou Cotile, with the Federals falling back and burning every house in the neighborhood. On April 29 the Federals fell back to five miles beyond McNutt's Hill, where a desperate fight took place, in which the 7th was very actively engaged. Major's division, in which the 7th was a part, was sent to Cheneyville, with orders to strike the Federal rear on the Red River and annoy his transports and gunboats, and if possible, cut off the Federals' communications with the Mississippi River. On May 6, the 7th regiment and the rest of the brigade, now under the command of recently made brigadier general, William P. Hardeman, was ordered to reinforce General Bee at Lecompte, and made a forced march of twenty-nine miles to that place. The battle here was a small, but brilliant affair, and during the night the Federals retreated to Alexandria, leaving a small force to protect their rear.

On May 15 the 7th regiment formed a line of battle along with the 4th and 5th regiments in the edge of Choctaw Swamp near the

Marksville prairie. They managed to check the advancing Federals and spent the night bivouacked in the edge of the prairie.

The Battle of Mansura occurred on May 16, was primarily an artillery duel, but the 7th, along with the 4th and 5th, occupied the right wing of the Confederate line and in order to draw off the Federal artillery from the center of the Confederate line, made a daring gallop down into the valley between the two lines of battle, with forty-two Federal cannons roaring at them. As they passed in front of the Confederate forces, the artillery moved to the rear, and Terrell's cavalry joined the three regiments across the battlefield. The 7th, 4th, and 5th were then left as rear guard to the Confederates and had to meet a furious charge of the Federals, who by this time had figured out the meaning of the unusual cavalry movement by the 4th, 5th, and 7th. The three Texas regiments, armed with their Enfield rifles, repulsed the Federals with numerous casualties, saving the Confederate artillery.

The 7th Texas Cavalry fought again in the battle of Yellow Bayou, which was the last major affair of the Red River Campaign, and on June 23, with Brigadier General Hardeman still in command of the brigade of which the 7th was a part, they were ordered to the little town of Trinity, on the Black River. After a march of about a week, the brigade arrived at their destination and made camp. Here the 5th regiment was stationed, and the 4th regiment was sent down river to Johnson's Ferry to observe the movements of the Federal forces in that region. The 7th regiment moved to Liddell's plantation, a few miles below Trinity. Two raids were planned and carried out while the brigade was on the Black River, one on Vidalia, a Federal outpost opposite Natchez, and the other a raid on the Mississippi River, where they captured 1,000 sacks of corn.

The camps on the Black River were abandoned on August 29, and the brigade moved up the Tensas and over to Waterproof on the Mississippi River. On September 1, the brigade started for Arkansas, and upon arriving there, spent most of their time marching, countermarching, drilling, and in policing camp. On September 11, the brigade crossed the Arkansas River fifteen miles above Arkansas Post and marched up on the opposite side to within a few miles of Pine Bluff. After much marching and countermarching, the brigade returned to their old camps at Rusk Lake, near Washington, Arkansas, where they remained until October 27. About this time the brigade was ordered to Fulton, where they engaged in drill and fatigue duty until ordered to Texas about November 7. Preparations were made and the regiments departed individually, reuniting at White Oak Shoals on about December 6. They then moved to Nacogdoches for awhile, but soon moved again to Alto, where they remained during the month of December, the regiments being camped in the vicinity. From there the regiments moved to Crockett in Houston County, and then to Hall's Bluff, where they established their headquarters about January 16, 1865. They remained here until March 11, 1865, when they moved to Fairfield in Freestone County. From here they moved to Alto Springs in Falls County, and on March 21, to Millican in Grimes County. During all this time, since December 4, 1864, the brigade had been commanded by Colonels Hampton and Waller, alternately, but on March 26, Brigadier General Hardeman returned to the brigade and took command.[13]

Much reorganization was now taking place in the Texas commands, and the brigade of which the 7th was a part (Green's old brigade), was broken up. The 4th regiment was retained for General Hardeman's new brigade, but the 5th was transferred to Debray's brigade, and the 7th was transferred to Terrell's brigade back in Louisiana, replacing Likens' regiment of that brigade, which had been ordered dismounted.[14]

Now a member of Terrell's brigade of mounted cavalry, the 7th regiment spent the remainder of their time in Louisiana in operations against jayhawkers, stragglers, and deserters.[15]

Some time between April 21 and May 1, 1865, the 7th Texas Cavalry broke camp near Grand Ecore, along with the 1st Texas Cavalry, commanded by Colonel Yager, and the 34th Texas Cavalry, commanded by Lt. Colonel Robertson (Terrell's old regiment). They headed for Texas again, back through Pleasant Hill and Mansfield, Shreveport, and Marshall, Texas. Once through Marshall, the brigade went on through Tyler and Athens, making their last camp together at Wild Cat Bluff on the Trinity River, where a ferry operated on the route to the west. The brigade received the news on about May 14, 1865, of the surrender of General Lee, while still in camp at the Wild Cat crossing, waiting for Colonel Terrell to arrive from Marshall. Before Terrell arrived, however, the regimental commanders had apparently assessed the situation, and did as most of the other commands west of the Mississippi River were doing at this time; they called the men together and told them to go home.[16]

FOOTNOTES

[1]Henderson, *Texas in the Confederacy,* 71-78.

[2]Evans, *Confederate Military History,* Vol. XI, 199-200.

[3]Henderson, *Texas in the Confederacy,* 94-96.

[4]Taylor, *Destruction and Reconstruction,* 189.

[5]Edgar, *History of De Bray's (26th) Regiment of Texas Cavalry,* 10-11; Taylor, *Destruction and Reconstruction,* 186.

[6]Smith & Mullins, eds., "The Diary of H.C. Medford, Confederate Soldier."

[7]Plummer, *Confederate Victory at Mansfield,* 21, 28-29.

[8]Taylor, *Destruction and Reconstruction,* 201; Johnson, *Red River Campaign,* 155; Plummer, *Confederate Victory at Mansfield,* 34.

[9]Plummer, *Confederate Victory at Mansfield,* 34.

[10]Taylor, *Destruction and Reconstruction,* 214-216.

[11]Johnson, *Red River Campaign,* 220.

[12]*Official Records,* XXXIV, 611-614.

[13]Wooten, *Comprehensive History of Texas,* 733-738.

[14]*Official Records,* XLVIII, 1390.

[15]Winters, *Civil War in Louisiana,* 416.

[16]Terrell, *From Texas to Mexico in 1865,* 1-5.

Appendix VII

Herbert's Texas Cavalry
(7th Texas Cavalry)

FIELD AND STAFF

Steele, William, Colonel. Promoted to brigadier general.
Sutton, John S., Lt. Colonel.
Bagby, Arthur P., Major, Colonel. Promoted to brigadier general on March 17, 1864.
Herbert, P.T., Lt. Colonel. Colonel may have been replaced by Hoffman as commander late in the war.
Hoffman, Gustave, Major, Colonel. May have taken over command of the regiment late in the war.
Jordan, Powhatan, Lt. Colonel.
Cupples, George, Surgeon.
Felton, R., Surgeon.
Cunningham, J.W., Assistant Surgeon.
Greenwood, Thomas B., Assistant Surgeon. Surgeon.
Hunter, Henry J., Assistant Surgeon.
Ogden, M.L., Assistant Quartermaster.
Fisher, L.C., Assistant Quartermaster.
Broyles, B.F., Assistant Quartermaster, 1st Lieutenant. Captured near Franklin, La., on April 13, 1863. Was wounded in the hand in the battle for Galveston, Texas, on January 1, 1863.[2] Broyles was from Palestine, Texas, in Anderson County. He was taken prisoner near Franklin, Louisiana, was shipped aboard the *Maple Leaf,* a Federal prison ship, to the East. However, he and the other prisoners managed to take over the ship, run it aground, and make their way to Richmond. He made his way back to Texas and rejoined the 7th Texas, remaining with it and later Terrell's brigade until it was disbanded at Wild Cat Bluff on the Trinity River.[3]
Lee, A.K., A.C.S.
Howard, W., may have been adjutant.
McIvers, W.W., Surgeon.
McLean, H.R., Captain, A.C.S.

COMPANY A

This company was successively designated as Captain Jordan's Company, 3rd Regiment, Sibley's Brigade, Texas Mounted Volunteers; Captain Thurmond's Company; and Company A, 7th Regiment Texas Mounted Volunteers; and Company A, 7th Regiment Texas Cavalry.

Jordan, Powhatan, Captain. Promoted to Lt. Colonel.
Thurmond, Alfred S., 1st Lieutenant, Captain.
Burke, James A., 2nd Lieutenant. Resigned January 11, 1865.
McGrew, Henry H., 2nd and 1st Lieutenant.
Dowden, J. A., and Lieutenant. Resigned February 25, 1865.

COMPANY B

This company was successively designated as Captain Hoffman's Company, 6th Regiment Texas Mounted Volunteers; Captain Hoffman's Company, 7th Regiment Texas Mounted Volunteers; and Company B, 7th Regiment Texas Cavalry.

Hoffman, Gustav, Captain, Major, Colonel.
Schwarzhoff, Scipio, 1st Lieutenant, Captain.
Conrad, Carl, 2nd and 1st Lieutenant.
Weichold, H., 2nd Lieutenant.
Eggeling, Julius, 2nd Lieutenant. Resigned May, 1864.

COMPANY C

This company was successively designated as Captain Burrows' Company, 7th Regiment Texas Mounted Volunteers; Captain Burrows' Company, 3rd Regiment Sibley's Brigade Mounted Volunteers; and Company C 7th Regiment Texas Cavalry.

Burrows, Hiram M., Captain. Resigned.
Robinson, John C., 1st Lieutenant, Captain.
Snyder, John W., 2nd and 1st Lieutenant.
Lewis, James F., 2nd Lieutenant.

COMPANY D
''Angelina Troop''

Successively designated as Captain Cleaver's Company; Captain Cleaver's Company, 3rd Regiment, Sibley's Brigade, Texas Mounted Volunteers; Captain Cleaver's Company; and Company D, 7th Regiment Texas Mounted Volunteers; and Company D, 7th Regiment Texas Cavalry.

Cleaver, William H., Captain, Resigned July 1, 1862.
Hudeburgh, A.C., 1st Lieutenant. Resigned April 20, 1862.
Parton, H.A., 2nd and 1st Lieutenant, Captain.
Eaton, G.W., 2nd Lieutenant. Resigned June 14, 1862.
Thompson, R.W., 2nd and 1st Lieutenant.
Fuller, B.J., 2nd Lieutenant.
Cushman, M., 2nd Lieutenant.
Brown, Sam, 2nd Lieutenant.

COMPANY E
"Trinity County Cavalry"

This company was known at various times as Captain Kirksey's Company; Captain Kirksey's Company, 3rd Regiment, Sibley's Brigade, Texas Mounted Volunteers; Company E, 7th Regiment, Texas Mounted Volunteers; and Company E, 7th Regiment Texas Cavalry.

Kirksey, W.L. Captain. Resigned February 25, 1865.
Tullos, S.F., 1st Lieutenant. Resigned February 17, 1862.
Chandler, C.T., 2nd Lieutenant. Died January, 1862.
Heslep, J.O., 2nd and 1st Lieutenant. Cashiered on August 16, 1864. No reason shown.
Oglesby, J.C., 2nd Lieutenant. Cashiered August 16, 1864. No reason given.
McLendon, H.M., 2nd Lieutenant.
Farley, J.A., 2nd Lieutenant.

COMPANY F
"New Salem Invincibles"

Successively designated as Captain Wiggins' Company of Cavalry; Captain Wiggins' Company, 3rd Regiment, Sibley's Brigade, Texas Mounted Volunteers; Captain Wiggins' Company; and Company F, 7th Regiment, Texas Mounted Volunteers; and Company F, 7th Regiment Texas Cavalry.

Wiggins, James F., Captain.
Gray, J.W., 1st Lieutenant. Died March 27, 1862.
Goldsberry, A.P., 2nd Lieutenant. Resigned November 14, 1861.
Wiggins, W.C., 2nd Lieutenant. Resigned January 1, 1863.
Cook, C.W., 1st Lieutenant.
Montgomery, S.D., 2nd Lieutenant.
Edmons, T.H., 2nd Lieutenant.
Johnson, P.H., 2nd Lieutenant.

COMPANY G

This company was successively designated as Captain Fisher's Company, 3rd Regiment, Sibley's Brigade Mounted Volunteers; Captain Fisher's Company, 7th Regiment Texas Mounted Volunteers; and Company G, 7th Regiment Texas Cavalry.

Fisher, H.W., Captain.
Holiday, William, 1st Lieutenant.
Lee, W.H., 2nd Lieutenant. Resigned September 13, 1862.
Scott, John C., 2nd Lieutenant. Resigned November 2, 1861.

Middleton, A., 2nd Lieutenant. Resigned February 28, 1862. Died March, 1862.
Robbin, W., 2nd Lieutenant.
Sessums, B.D., 2nd Lieutenant. Resigned May 30, 1864.

COMPANY H

This company was successively designated as Captain Adair's Company; and Captain Haley's Company, 3rd Regiment, Sibley's Brigade Texas Mounted Volunteers; Company H, 7th Regiment Texas Mounted Volunteers; and Company H, 7th Regiment Texas Cavalry.

Adair, Isaac, Captain. Died April 9, 1862.
Haley, Charles Q., 1st Lieutenant, Captain. Resigned July 2, 1863.
Arrington, B.B., 2nd and 1st Lieutenant. Captain. Enlisted October 5, 1861, at Crockett, Houston County, Texas. In 1862 Arrington was 28 years old.
Daniel, James, 2nd and 1st Lieutenant.
Hennes, George W., 2nd Lieutenant.
Davis, W.W., 2nd Lieutenant.
Porter, J.M., 2nd Lieutenant.

COMPANY I
"Anderson County Company of Cavalry"

This company was successively designated as Captain Gardner's Company of Cavalry; Captain Gardner's Company, 3rd Regiment, Sibley's Brigade Mounted Volunteers; Company I, 7th Regiment Texas Mounted Volunteers; and Company I, 7th Regiment Texas Cavalry.

Gardner, James W., Captain. Died June 28, 1862.
Key, William B., 1st Lieutenant. Resigned February 10, 1862.
Taylor, John W., 2nd Lieutenant. Captain.
Mills, Charles H., 2nd Lieutenant. Resigned October 4, 1862.
Horn, C.C., 1st Lieutenant. Captain.
Broyles, B.F., 2nd and 1st Lieutenant. Made Brigade Assistant Quartermaster (See Field and Staff).
Alexander, J.H., 2nd Lieutenant.
Bishop, J.E., 2nd Lieutenant.
Lumpkin, Wilson, 2nd Lieutenant.

COMPANY K

This company successively designated as Captain Moody's Company, 3rd Regiment, Sibley's Brigade Texas Mounted Volunteers; Company K, 7th Regiment Texas Mounted Men; and Company K, 7th Regiment Texas Cavalry.

Moody, Thomas O., Captain. Resigned September 2, 1863.
Steele, L. G. A., 1st Lieutenant. Resigned October 24, 1863.
Bowman, Isaac G., 2nd Lieutenant. Captain.
Smith, John P., 2nd and 1st Lieutenant. Promoted to Major as I.A.A.G. in Bagby's Division of Cavalry.
Dagget, E. B., 2nd Lieutenant.

FOOTNOTES

[1] Microfilm 323, *Herbert's (7th) Texas Cavalry,* Roll 45.
[2] Microfilm 323, *Herbert's (7th) Texas Cavalry,* Roll 45.
[3] *The Lone Star State,* Lewis Publishing Company.

APPENDIX VIII

Surrender Terms of the Trans-Mississippi Department[1]

GENERAL ORDERS, NO. 61 — HDQRS, MIL. DIV. OF WEST MISSISSIPPI, New Orleans, La., May 26, 1865.

1. By the terms of a convention entered into this day, on the part of General E.K. Smith, commanding the Trans-Mississippi Department and Maj. Gen. E.R.S. Canby, commanding the Military Division of West Mississippi, the forces, military and naval, of the Trans-Mississippi Department, and the public property under their control, have been surrendered to the authorities of the United States.

In carrying out the stipulations of this convention the following conditions will be observed:

1. All acts of hostility on the part of both armies are to cease from this date.
2. The officers and men of the Confederate Army and Navy within the limits of the Trans-Mississippi Department, to be paroled until duly exchanged or otherwise released from the obligations of their parole by the authority of the Government of the United States. Duplicate rolls of all officers, and men paroled to be retained by such officers as may be designated by the parties to this convention, officers given their individual paroles and commanders of regiments, battalions, companies, or detachments signing a like parole for the men of their respective commands.
3. Artillery, small arms, ammunition, and other property of the Confederate Government, including gunboats and transports, to be turned over to the officers appointed to receive them on the part of the Government of the United States. Duplicate inventories retained by the officer delivering and the other by the officer receiving it, for the information of their respective commanders.
4. The officers and men paroled under this agreement to be permitted to return to their homes, with the assurance that they will not be disturbed by the authorities of the United Staes so long as they continue to observe the conditions of paroles and the laws in force where they reside. Persons, residents of Northern States, and not excepted in the amnesty proclamation of the President, will be permitted to return to their homes on taking the oath of allegiance to the United States. See paragraph II. General Orders, No. 55, current series.*

5. The surrender of property will not include the side arms or private horses or baggage of officers, nor the horses which are, in good faith, the private property of enlisted men. These last will be allowed to take their horses to their homes to be used for private purposes only.

6. The time, mode, and place of paroling and surrender of property will be fixed by the respective commanders, and will be carried out by commissioners appointed by them.

7. The terms and conditions of this convention to extend to all officers and men of the Army and Navy of the Confederate States, being in or belonging to the Trans-Mississippi Department.

8. The troops and property to be surrendered within the limits of the Division of Missouri will be turned over to commissioners appointed by the commander of that division, and the men and material of the Navy to commissioners appointed by the commanders of the Mississippi and West Gulf Squadrons, respectively, according to the limits in which the said men and material may be found.

9. Transportation and subsistence to be furnished at public cost to the officers and men (after being paroled) to the nearest practicable point to their homes.

10. If the U.S. troops designated for the garrison of interior points should not reach their destinations before the work of paroling is completed, suitable guards will be detailed for the protection of the public property. These guards, when relieved, will surrender their arms and be paroled in accordance with the terms of this convention.

II. The U.S. troops sent into the interior of the country will be kept well in hand, in a state of the most exact discipline, and in constant readiness for any service which they may be called upon to perform. When detachments are made for the purpose of protecting the inhabitants against jayhawkers and other lawless characters, and on all marches through the country, the conduct of officers and men must be such as to inspire the people with confidence and respect, and no depredations, however slight, or interference with the citizens in their lawful pursuits will be permitted.

III. To guard against the waste or loss of public property not under the control of the Confederate military and naval authorities, the civil officers or agents in charge of such until relieved by the proper officers or agents of that Government. All sales of such property, or transfers, except to the authorized agents of the Government, are forbidden, and any attempt to conceal or withhold it will work the immediate

forfeiture of any private interest that may be involved.
IV. Private property will not be interfered with unless required for "public use," and where this is necessary it will be taken in an orderly and regular manner, under the orders of the commanding officer, and the proper receipts will be given. Property so received will be disposed of and accounted for as any other public property.
V. 1. Until the commercial restrictions and the blockade of the Gulf ports are removed by the President no foreign or general commerce with those ports or with the interior of the country west of the Mississippi (within the limits of this division) can be permitted, and trade will be limited to the wants of the Army and Navy and the necessities of the inhabitants within the limits of military occupation. To the extent of these necessities military permits and clearances may be given for supplies not prohibited by existing orders, but no permit or clearance will be given to any point that is not occupied by a military or naval force.

2. In the neighborhood of military posts the inhabitants may freely bring in their produce and take out such supplies as may be required for plantation and family use. Live-stock, provisions of all kinds, fuel, and other products and material required by the Army and Navy, or for the use of the inhabitants, may be freely sold in open market; but no other products of insurrectionary districts can be sold or shipped, except by delivery to the quartermaster's department for consignment to a purchasing agent of the Treasury Department.

3. No trade stores or trade permits for the interior will be permitted or recognized until the regulations of the Treasury Department can be extended over the country to be occupied, and until then no clearances or permits will be granted for any point that is not occupied by the troops of the United States.

VI. Under the authority of the Executive order of April 29, 1865, all "well-disposed persons" who accept in good faith the President's invitation "to return to peaceful pursuits" are assured that they may resume their usual avocations, not only without molestation, but, if necessary, under the protection of the U.S. troops, conforming to the regulations of the Treasury Department, and to the additional condition of not fabricating or dealing in articles contraband of war.

By order of Maj. Gen. E.R.S. Canby:

C.T. CHRISTENSEN,
Lieutenant-Colonel and Assistant Adjutant-General.

[1] *Official Records,* XLVIII, Part 2, 604-606.

Appendix IX

Appeals by Generals E. Kirby Smith and J.B. Magruder
To the Army and People of the Trans-Mississippi Department[1]

Headquarters Trans-Mississippi Department,
Shreveport, La., April 21, 1865.

Soldiers of the Trans-Mississippi Army:

The crisis of our revolution is at hand. Great disasters have overtaken us. The Army of Northern Virginia and our Commander-in-Chief are prisoners of war. With you rests the hopes of our nation, and upon your action depends the fate of our people. I appeal to you in the name of the cause you have so heroically maintained — in the name of your firesides and families so dear to you — in the name of your bleeding country, whose future is in your hands. Show that you are worthy of your position in history. Prove to the world that your hearts have not failed in the hour of disaster, and that at the last moment you will sustain the holy cause which has been so gloriously battled for by your brethren east of the Mississippi.

You possess the means of long-resisting invasion. You have hopes of succor from abroad — protract the struggle and you will surely receive the aid of nations who already deeply sympathize with you.

Stand by your colors — maintain your discipline. The great resources of this department, its vast extent, the numbers, the discipline, and the efficiency of the army, will secure to our country terms that a proud people can with honor accept, and may, under the Providence of God, be the means of checking the triumph of our enemy and of securing the final success of our cause.

E. KIRBY SMITH
General.

GENERAL ORDERS, No. 20

HEADQUARTERS DISTRICT OF TEXAS, NEW MEXICO, AND ARIZONA,
Houston, April 23, 1865.

The major-general, commanding the District of Texas, New Mexico, and Arizona, deems it proper, in view of recent events, to call upon the army and patriotic citizens of Texas to set an example of devotion, bravery, and patriotism, worthy of the holy cause of liberty and independence, and of the great efforts heretofore made by the army and the people of Texas to advocate and uphold it. The enemy threatens our coast and will bring his great undivided resources for a successful invasion of the State. Let him be met with unanimity and Spartan courage, and he will be unsuccessful, as he has been in Texas. Let him be met at the water's edge, and let him pay dearly for every inch of territory he may acquire. Six hundred Frenchmen under the First Napoleon recaptured France from her enemies. Forty-two Irish soldiers, on our own soil, drove 15,000 men to sea. The Army of the Trans-Mississippi Department is larger, in finer order, and better supplied than ever. There are no navigable streams in Texas, therefore the enemy will be divested of the great power of steam, which he has elsewhere relied upon. Crops have been bountiful; our armies can therefore be supplied in almost any part of Texas. There is no reason for despondency, and if the people of Texas will it, they can successfully defend their territory for an indefinite period. The major-general commanding therefore exhorts the soldiers of the army to stand firmly by their colors, and obey the orders of their officers, and recommends to the citizens that they devote themselves still more fully to the cultivation of breadstuffs; for should our armies be unsuccessful in the east, every gallant soldier will rally to the banner of the Confederacy, which will still float defiantly west of the Mississippi River.

By command of Maj. Gen. J. B. Magruder:

EDMUND P. TURNER,
Captain and Assistant Adjutant-General.

BIBLIOGRAPHY

PRIMARY SOURCES

U.S. GOVERNMENT DOCUMENTS AND PAPERS

Microfilm Copy No. 323, The National Archives, National Archives and Records Service, General Services Administration, Washington, D.C., 1961: *Herbert's (7th) Texas Cavalry,* Roll 45. *Likens' (35th) Texas Cavalry,* Rolls 171 and 177. *Terrell's (37th) Texas Cavalry,* Rolls 177, 178, and 179. *Yager's (1st) Texas Cavalry,* Rolls 5 and 7.

Microfilm specially prepared by the National Archives, National Archives and Records Service, General Services Administration, Washington D.C., March, 1976. Covers Muster Rolls and other papers relating to Terrell's Texas Cavalry Regiment. Prepared at the expense of the author and in his possession.

United States War Department, *War of the Rebellion: A Compilation of the Official Records of the Union and Confederate Armies,* 70 volumes in 128, Washington, Government Printing Office, 1890-1901.

BOOKS

Barr, Alwyn, *Polignac's Texas Brigade,* Texas Gulf Coast Historical Association Publication, Vol. VIII, No. 1, Nov., 1964.

Bitton, Davis, editor, *The Reminiscences and Civil War Letters of Levi Lamoni Wight,* Salt Lake City, University of Utah Press, 1970.

Blessington, Joseph P., *The Campaigns of Walker's Texas Division,* New York, Lange, Little and Co., 1875.

Edgar, Thomas H., compiler, *History of De Bray's (26th) Regiment of Texas Cavalry,* Galveston, Press of A.A. Finck and Co., 1898.

Evans, Clement A., *Confederate Military History,* Atlanta, Confederate Publishing Company, 1899, Vols. X & XI.

Faulk, Odie, *Tom Green, Fightin' Texan,* Waco, Texian Press, 1963.

Fitzhugh, Lester N., *Texas Batteries, Battalions, Regiments, Commanders and Field Officers, Confederate States Army, 1861-1865,* Midlothian, Texas, Mirror Press, 1959.

Henderson, Harry McCorry, *Texas in the Confederacy,* San Antonio, The Naylor Company, 1955.

Johnson, Ludwell H., *Red River Campaign,* Baltimore, The John Hopkins Press, 1958.

Johnson, Sidney S., *Texans Who Wore the Gray,* Tyler, Texas, 1907.

Plummer, Alonzo H., *Confederate Victory At Mansfield,* Mansfield, Louisiana, Ideal Printing Company, 1969.
Taylor, Richard, *Destruction and Reconstruction,* edited by Richard B. Harwell, New York, Longmans, Green, and Company, 1955.
Terrell, Alexander W., *From Texas to Mexico and the Court of Maximilian in 1865,* edited by Fannie E. Ratchford, Dallas, Texas Book Club, 1933.
Vaughn, Michael J., *The History of Cayuga and Cross Roads, Texas,* Waco, Texas, Texian Press, 1967.
Weddle, Robert S., *Plow Horse Cavalry,* Austin, Madrona Press, 1974.
Winters, John D., *The Civil War in Louisiana,* Baton Rouge, Louisiana State University Press, 1963.
Wright, Marcus J., Brig. General, *Texas in the War, 1861-1865,* edited and notes by Colonel Harold B. Simpson, Hillsboro: Hill Jr. College Press, 1965.
Wooten, Dudley G., editor, *A Comprehensive History of Texas 1865-1897,* 2 Volumes, Dallas, William G. Scarff, 1898.

DIARIES AND JOURNALS

Barr, Alwyn, editor, "William T. Mechling's Journal of the Red River Campaign, April 7-May 10, 1864," *Texana,* I, No. 4, Fall, 1963.
Smith, Rebecca W. and Mullins, Marion, editors, "The Diary of H.C. Medford, Confederate Soldier, 1864," *Southwestern Historical Quarterly,* XXXIV, January, 1931.

ARTICLES

Bee, Hamilton P., "Battle of Pleasant Hill — An Error Corrected," Southern Historical Society Papers, VIII, 1880.
Hewitt, J.E., "The Battle of Mansfield, La.", *Confederate Veteran,* XXXIII, May, 1925.
"Judge Alexander Watkins Terrell," *Confederate Veteran,* XX, Dec., 1912.
Meiners, Fredericka, "Hamilton P. Bee in the Red River Campaign," *Southwestern Historical Quarterly,* LXXVIII, No. 1, July, 1974.
Moreland, Sinclair, "Life Sketch of A.W. Terrell," *The Home and State,* Dallas, Vol. 14, No. 12, Sept., 1912.
Sliger, J.E., "How General Taylor Fought the Battle of Mansfield, La.," *Confederate Veteran,* XXXI, December, 1923.
"The Battle of Yellow Bayou," *Confederate Veteran,* XXV, Feb., 1917.

POEMS

"Terrell's Poem about the death of John Wilkes Booth," *Confederate Veteran,* April, 1913.

THESIS

Chamberlain, Charles K., *Alexander Watkins Terrell, Citizen Statesman,* Ph.D. dissertation, University of Texas, 1956.

NEWSPAPERS

The Austin *Statesman,* September 10, 1912.
The Dallas *Times Herald,* October 1, 1864.
The Galveston *Tri-Weekly News,* April 25, 1864.
The Houston *Daily-Telegraph,* April 24, 1864 and June 8, 1864.

PAMPHLETS

Texas Civil War Centennial Program for the Centennial Commemoration of the Red River Campaign, Odessa, Texas, West Texas Office Supply, A Publication of the Texas Civil War Centennial Commission and Texas Historical Survey Committee, 1962. Barr, Alwyn, "Texas Losses in the Red River Campaign, 1864."

Souvenir History Program for the Centennial Commemoration of the Red River Campaign, Texas State Historical Survey Committee, 1964, Fitzhugh, Lester N., "Texas Forces in the Red River Campaign, March — May, 1864;" Smith, E. Kirby, General, "Excerpts from the Official Report of the Red River Campaign."

Stephens, Robert W., *August Buchel, Texas Soldier of Fortune,* Privately printed booklet by Mr. Stephens, who resides in Dallas, Texas. Much of the information in Mr. Stephen's booklet comes from unpublished material furnished to him by Mr. August R. Buchel of Dallas, great-grandnephew of Colonel Buchel. Also included in the booklet is information by Louis Lenz of Houston, who had an extensive collection of Buchel's personal papers.

UNPUBLISHED MANUSCRIPTS

Horn, Margaret, *Introduction to the History of Hunt County,* Greenville, Texas, (No date); this manuscript is on file in the W. Walworth Harrison Public Library, Greenville, Texas.

SCRAPBOOKS

"Account of the Battle of Blair's Landing," *Tom Green Scrapbook,* Austin, Texas, Texas State Library, Archives Division.

Baker, D.W.C., *Texas Scrapbook,* N.Y., A.S. Banners & Co., 1875, Austin, Texas, Texas State Library, Archives Division.

Riddell, Sarah Glenn, *Scrapbook, 1864,* (Tom Green's sister-in-law), Texas State Library, Archives Division.

ARCHIVES

Alexander Watkins Terrell Papers, Texas State Library, Archives Division, Austin, Texas.

Alexander Watkins Terrell Papers, University Archives, The Eugene C. Barker Texas History Center, University of Texas at Austin, Austin, Texas.

Terrell's Texas Cavalry Regiment (37th), National Archives, Washington, D.C.

SECONDARY SOURCES

ARTICLES

Debray, Xavier B., "A Sketch of Debray's Twenty-sixth Regiment of Texas Cavalry," *Southern Historical Society Papers,* XIII, 1885.

Muir, Andrew Forest, "Dick Dowling and the Battle of Sabine Pass," *Civil War History,* IV, No. 4.

BOOKS

Acheson, Sam and O'Connell, Julie Ann Hudson, editors, *George Washington Diamond's Account of the Great Hanging at Gainesville,* 1862, Austin, The Texas State Historical Association, 1963.

Bearss, Edwin C., editor, *A Louisiana Confederate, Diary of Felix Pierre Poche',* Natchitoches, Louisiana, Northwestern State University, 1972.

Baker, Eugene C., Potts, Charles Shirley, and Rausdell, Charles W., *A School History of Texas,* Chicago, Row, Peterson and Co., 1913.

Booth, Andrew B., compiler, *Records of Louisiana Confederate Soldiers and Louisiana Commands,* Vol. III, Book 2, New Orleans, 1920.

Current, Richard N., Williams, T. Harry, Freidel, Frank, *American History,* New York, Alfred A. Knopf, 1961.

Daniel, L.E., *Personnel of the Texas State Government,* San Antonio, Maverick Printing House, 1892.
Printing House, 1892.

Heartsill, W.W., *Fourteen Hundred and 91 Days in the Confederate Army,* edited by Bell Irvin Wiley, Jackson, Tennessee, McCowat-Mercer Press, 1953.

Henry, Robert Selph, *the Story of the Confederacy,* New York, Grosset & Dunlap, 1931.

Johnson, Frank W., *A History of Texas and Texans,* Chicago & N.Y., The American Historical Society, 1914.

Lane, Walter P., *Adventures and Recollections of General Walter P. Lane,* Austin, Pemberton, 1970.

Lubbock, Francis R., *Six Decades In Texas,* Austin, The Pemberton Press, 1968.

Newton, Lewis W. and Gambrell, Herbert P., *A Social and Political History of Texas,* Dallas, The Southwest Press, 1932.

Oates, Stephen B., *Confederate Cavalry West of the River,* Austin, University of Texas Press, 1961.

Pennington, L. A., Hough, Romeyn B. Jr., and Case, H. W., *The Psychology of Military Leadership,* New York, Prentice-Hall Inc., 1943.

Flinn, Frank M., *Campaigning With Banks in Louisiana, '63 and '64,* Lynn, Mass., Press of Thomas P. Nichols, 1887.

The Lone Star State, Lewis Publishing Co., 1893.

Wakelyn, Jon L., *Biographical Dictionary of the Confederacy,* Westport, Connecticut, Greenwood Press, 1977.

Warenskjold, Elise, *The Lady With the Pen,* edited by C. A. Clausen, Norwegian-American Historical Association, Northfield, Minnesota, 1961.

Yeary, Mamie, *Reminiscences of the Boys in Gray, 1861-1865,* Dallas, Smith & Lamer Publishing House, 1912.

Zuber, William Physick, *My Eighty Years in Texas,* edited by Janis Boyle Mayfield, Austin, University of Texas Press, 1971.

CEMETERY RECORDS

Card Index of Graves of Civil War Soldiers in Navarro County, Corsicana Public Library, Corsicana, Texas.

DIARIES

Barr, Alwyn, Editor, "The Civil War Diary of James Allen Hamilton, 1861-1864, *Texana,* II, No. 2, Summer, 1964.

ENCYCLOPEDIAS

The Encyclopedia Americana, 1956 Edition, New York, Americana Corporation, 1956.

STATE GOVERNMENT PUBLICATIONS

Annual Report of the Adjutant General of the State of Illinois, Springfield, Baker and Phillips, Printers, 1863.

U.S. GOVERNMENT PUBLICATIONS

Department of the Army, *American Military History,* 1607-1953, July, 1956.
Microfilm Copy No. 323, The National Archives, National Archives and Records Service, General Services Administration, Washington, D.C., 1961: *Eighth Texas Cavalry,* Roll 51.
Steele, Matthew Forney, *American Campaigns,* Vol. I, Washington, United States Infantry Association, 1943.

MUSTER ROLLS

Company B, 32nd Texas Cavalry, Texas State Library, Archives Division, Austin, Texas.

PAMPHLETS

Manucy, Albert, *Artillery Through the Ages,* Washington: United States Government Printing Office, 1949.
Price, William H., *The Civil War Handbook,* Fairfax, Virginia, Prince Lithograph Co., Inc., 1961.
Texas Civil War Centennial Commission and Texas Historical Survey Committee, *Texas in the Civil War:* A Resume History, Odessa, Texas, West Texas Office Supply, 1962.

PERSONAL INTERVIEWS

Mr. Robert Hood, Mrs. Adie Simmons, and Mrs. Theda Spencer, all of Breckenridge, Texas, and Mrs. Dorothy Dennison, Amarillo, Texas, all grandchildren of Alexander E. Hulse.

INDEX

A

Abdul Hamid, 87
Albuquerque, N.M., 160
Alexandria, La., 8, 10, 39-43, 49, 50, 156, 163
Allen, Gov. ____ (La.), 58
Alto, Tex., 165
Alto Springs (Falls Co.), Tex., 165
Arkansas Post, Ark., 164
Arkansas River, 164
Armant, Col. ____, 16, 19
Atchafalaya (Bayou) River (La.), 44-49, 156
Athens, Tex., 54, 55, 165
Austin, Tex., 52, 53, 55, 59, 82, 85, 88

B

Bagby, Arthur Pendleton, 12, 18, 21, 31, 46, 70-71, 160-162
Bankhead, Brig. Gen. ____, 3
Banks, Nathaniel Prentiss, 9-12, 27, 28, 35, 67, 162
Barbee, K.H., 45
Baylor, George W., 44
Bayou Boeuf (La.) 41, 42, 49, 156
Bayou Bourbeau (La.), 161
Bayou Cotile (La.), 163
Bayou Pierre (La.), 36, 39, 162
Bayou Rouge (La.), 48, 151
Bayou Teche (La.), 161
Beard, Col. ____, 16, 19
Beasley's Station, La., 40, 149, 156, 163
Bee, Hamilton Prioleau, 1, 8, 9, 14, 20, 21, 31, 33-36, 39-41, 61-62, 84, 148, 149, 155, 156, 162, 163
Benton, Lt. Col. ____, 37
Berwick Bay, La., 161
Black Hawk, 10, 37
Black River (La.), 164
Blair's Landing, La., 36, 149, 162, 163
Boca Chica (Tex.), 53
Bolivar Point, Tex., 147
Bonham, Tex., 2, 3, 6, 84
Boonville, Mo., 81
Booth, John Wilkes, 85, 86
Boston, Mass., 45
Brazos River (Tex.), 3, 85
Brisbin, Col. ____, 26
Brown, R.R., 7
Bryan, Guy M., 52
Buchel, Augustus Carl, 1, 9, 13-15, 17, 31, 33, 34, 46, 62-63, 75-76, 84, 88, 148, 155, 162
Buckner, Simon Bolivar, 46, 53, 60

C

Camp Bisland, La., 161
Camp Dixie, Tex., 8
Camp Groce, Tex., 3, 84
Camp Kelsoe Springs, Tex., 3
Camp Nelson, Ark., 83
Cam Sidney Johnston, Tex., 8
Camp Wharton, Tex., 7, 8
Canby, E.R.S., 172, 174
Cane River (La.), 39, 40, 163
Caney River (Tex.), 8, 148
Canfield, Maj. ____, 19
Caudle, Lt. Col. ____, 17
Chancellor, J.G., 4, 5, 7
Chapultepec, Mex., 59
Cheneyville, La., 163
Choctaw Swamp (La.), 163
Christensen, C.T., 174
Churchill, Thomas J., 28, 31, 32, 53, 73-74
Clack, Lt. Col. ____, 20
Clark, John B., 58
Cleveland, Grover, 87
Cloutierville, La., 39

Colorado River (Tex.), 148
Columbia, Tex., 8
Columbus, Tex., 3, 5, 7
Corsicana, Tex., 53
Corwin, Capt. ____, 12
Crockett, Tex., 165

D

Davis, E.J., 57
  Jefferson, 53, 55, 83
Day, Ballard A., 45
Debray, Xavier Blanchard, 7, 9, 12-15, 17, 31, 33-35, 42, 48, 62, 88
Douay, Gen. ____, 59
Duff, Col. ____, 57, 58, 84, 148

E

Elliott's Ferry, Tex., 148
Elmira, N.Y., 45, 46
Emory, Gen. ____, 18, 25, 26
Evergreen, La., 48

F

Fairfield, Tex., 165
Fayetteville, Tex., 8
Felton, James, L., 45, 46
Flinn, Frank M., 25
Flournoy, George, 55, 57, 59, 84
Ford, John Salmon (Rip), 52
Fort Bliss, Tex., 160
Fort Craig, N.M., 160
Fort Derussy, La., 10, 44
Fort Jessup, La., 11
Fort Worth, Tex., 81
Franklin, Gen. ____, 25
Fulton, Ark., 164

G

Galveston, Tex., 3, 4, 7, 8, 49, 55, 147, 148, 150, 161
Garcia, Col. ____, 57
Gentry, A.M., 147
Gilbreth, Mr. ____, 6
Gilley, Gabriel D., 45
Gonzales, Miss ____, 59
Gould, Nicholas C., 36, 66-67, 149
Grand Bayou Landing, La., 162, 163
Grand Ecore, La., 10-12, 35, 36, 39, 50, 52, 165
Green, Thomas (Tom), 1, 2, 9, 16, 18, 20, 21, 31, 34, 36-38, 44, 63, 149, 160, 161, 163
Gurney, Lt. Col. ____, 45

H

Halleck, Henry W., 10, 12
Hall's Bluff, Tex., 165
Hampton, Col. ____, 165
Hardeman, William Polk, 30, 38, 43, 55, 56, 59, 74, 84, 88, 160, 162, 163-165
Harrisburg, Tex., 150, 151
Havana, Cuba, 60
Hayden, Peyton R., 81
Hebert, P.O., 83, 147
Hempstead, Tex., 3, 50, 84, 160
Herbert, P.T., 161
Hill, Maj. ____, 55
Hindman, T.C., 57, 84
Hoffman, Gustave, 12
Houston, Sam, 82
Houston, Tex., 3, 7, 8, 44, 55, 85, 147, 151, 161
Howes, Brig. Gen. ____, 9
Hyllested, Maj. ____, 7

J

Jackson, A.M., 48
  Thomas Jonathan (Stonewall), 84
James, Constance, 88
Jeanningros, Gen. ____, 57
Johnson, ____, 49

Andrew, 59, 60
M.T., 55
Johnson's Ferry (La.), 164
Jones, A.C., 57
Mrs. Ann H. (Holliday), 85
G.W., 83
Jones Creek (Tex.), 7
Juarez, Benito, 53

K

Keatchie, La., 14, 28
King, W.H., 55, 57, 84

L

LaFourche, La., 161
Lamourie Bayou (La.), 42
Lane, Walter Paye, 20, 71, 84
Lavaca, Tex., 7
Lecompte, La., 42, 163
Lee, Albert, 14, 26
Robert E., 51, 52, 84, 165
Leona River (Tex.), 160
Likens, James B., 147, 148
Little Rock, Ark., 83
Logan's Ferry, La., 48
Logansport, La., 9, 50
Lubbock, Francis R., 82, 83

Mc

McCulloch, Henry E., 6, 7, 83
McKinney, Thomas F. (Tom), 55, 60
McNeil, P., 148
McNeill, Henry C., 30
McNutt's Hill, La., 40, 41, 163

M

Magruder, John Bankhead, 3, 6-9, 46, 57-59, 83, 84, 175, 176
Major, James P., 18, 20-22, 31, 41, 71, 156, 162
Mansfield, La., 1, 2, 9, 11-14, 16, 18, 19, 21, 22, 28-30, 33, 35, 36, 52, 84, 149, 155, 162, 165
Mansura, La., 43, 151, 164
Many, La., 11, 161
Marksville, La., 42, 43, 164
Marshall, Tex., 52-54, 165
Matagorda, Tex., 7
Mather's Bridge, La., 41
Maury, ____, 59
Maxey, Sam B., 150
Maximilian, 56, 59, 60, 84, 85
Mechling, William T., 34
Medford, H.C., 21
Mexico City, Mex., 58-60
Mier, Mex., 57, 58
Millican, Tex., 5, 165
Minden, La., 49
Mineral Wells, Tex., 88
Mississippi River, 163-165
Mitchell, Sallie D., 85
Mobile, Ala., 49
Monett's Ferry, La., 39, 40, 149, 156
Monterrey, Mex., 55, 57, 58, 84
Montgomery, La., 150
Moore, Gov. ____ (La.), 58
Moreauville, La., 156
Morgan, Hiram S., 2, 35
Morgan's Ferry, La., 45-48, 151, 156
Morganza, La., 46
Mouton, Alfred (Jean Jacques Alexandre), 16, 19, 22, 69, 84
Murrah, Pendleton, 9, 55, 56, 58
Murray, Caldean G., 3-7
Muscle Shoal Bayou (La.), 47

N

Nacogdoches, Tex., 165
Napoleon III, 59, 60, 84, 85
Natchez, Miss. 164
Natchitoches, La., 9, 10, 12, 39, 49, 161
Navasota, Tex., 84

New Braunfels, Tex., 1
New Orleans, La., 9, 10, 60
New York, N.Y., 60
Noble, Lt. Col. ____, 19

O

Onion Creek (Tex.), 55
Opelousas, La., 161
Osage, 37

P

Palmetto, Tex., 52
Parson, ____, 58
Parsons, Mosby, 28, 74
    William H., 36, 37, 76
Peach Orchard (La.), 155
Pine Bluff, Ark., 164
Pleasant Hill, La., 9, 11-14, 20, 28, 29, 31, 35, 36, 46, 52, 84, 149, 155, 161-163, 165
Polignac, Camille Armand Jules Marie, 18, 20, 21, 69-70
Preston, Gen. ____, 58
Price, Sterling, 28, 53, 58

Q

Quitman, Tex., 6

R

Randal, Horace, 18, 20, 21, 69
Ransom, Gen. ____, 25
Reager, George, 45, 46
Red River, 9, 10, 11, 36, 42, 44, 45, 47, 49, 50, 149, 163
Reynolds, Gov. ____ (Mo.), 58
Rice, John T., 45
Richland Creek (Tex.), 54
Richmond, Va., 45
Rio Grande River, 7, 53, 57
Roberts, Capt. ____, 55
Robertson, John C., 2-5, 7, 33, 54, 64, 157, 165
Roma, Tex., 57
Rusk Lake (Ark.), 164
Rusk, Tex., 9

S

Sabine Pass, Tex., 147
Sabine River, 9, 149
Sabinetown, Tex., 9, 12, 149
St. Joseph, Mo., 82
Salado Creek (Tex.), 160
Saluria, Tex., 148
San Antonio, Tex., 53, 59, 160
San Bernard, Tex., 148
San Luis Potosí, Mex., 59
Scurry, William Redi, 18, 20, 21, 83
Seven Mile Creek (La.), 21
Seward, William Henry, 10
Shelby, Joseph (Joe) Orville, 53
Sherman, William Tecumseh, 9, 10
Shreveport, La., 9-13, 28, 36, 52-54, 83, 151, 165
Sibley, Henry Hopkins, 160
Simmesport, La., 45, 46, 48, 156
Six Mile Bayou (La.), 1
Smith, A.J., 9, 10, 12, 29, 32, 49
    Ashbel, 148, 150
    Edmund Kirby, 3, 7, 9, 12, 50-55, 58, 64-65, 83, 172, 175
    Leon, 57
    Peter, 55
Springville, Tex., 5
Squires, Maj. ____, 53
Starr, Russell J., 5
Steele, Gen. ____, 10
    William, 160
Stone, Gen. ____, 26
Sutton, J. S., 160

T

Taylor, Richard, 1, 8, 9, 11, 12, 14-17, 19, 28, 35, 36, 61, 149, 155, 156, 161, 162
Ten Mile Bayou (La.), 1, 161, 162
Tensas River (La.), 164
Terrell, Alexander Watkins, 2, 3, 5, 7, 9, 13-16, 33, 35, 38, 43, 46-48, 52-60, 81-88. 149, 155-157, 163, 165
    Ann, 59, 82
    Christopher Joseph, 81
    Mrs. Christopher Joseph (Susan), 81
    James C., 81
    John J., 81
Texana, Tex., 7, 148
Trinity, La., 164
Trinity River (Tex.), 54, 157, 165
Tucker, Maj. ____, 5, 6
Turner, Edmund P., 57, 84, 147, 176
Tyler, Tex., 5-7, 147, 165

V

Valverde, N.M., 160
Velasco, Tex., 3, 148
Vera Cruz, Mex., 59, 60
Vicksburg, Miss., 9
Victoria, Tex., 148
Vidalia, La., 164
Virginia Point, Tex., 8

W

Waco, Tex., 3, 7, 8
Walker, Lt. Col. ____, 16, 19
    John G., 18, 20-22, 31, 57, 68, 84
Warren, James F., 35
Washington, Ark., 164
Washington, D.C., 10
Waterproof, La., 164
Watkins, Oscar, 57
Waul, Thomas N., 21
Webb, Lt. Col. ____, 14
West Barnard, Tex., 3
Wharton, John A., 44, 50
White Oak Shoals (Tex.), 165
Wight, Levi, 155, 156
Wilcox, Cadmus, 57, 84
Wild Cat Bluff, Tex., 54, 55, 79, 156-157, 165
Woods, Peter C., 36, 42, 66, 149

Y

Yager, William O., 156, 163, 165
Yellow Bayou, La., 44, 45, 151, 164

Z

Zuber, William, 45, 46